NFPA 70E®

Standard for

Electrical Safety in the Workplace®

2015 Edition

This edition of *NFPA 70E, Standard for Electrical Safety in the Workplace*, was prepared by the Technical Committee on Electrical Safety in the Workplace and released by the Correlating Committee on National Electrical Code®. It was issued by the Standards Council on July 14, 2014, with an effective date of July 29, 2014, and supersedes all previous editions.

A Tentative Interim Amendment (TIA) to 130.7(C)(10)(b)(1) and Tables H.3(a) and (b) was issued on August 14, 2014. For further information on tentative interim amendments, see Section 5 of the Regulations Governing the Development of NFPA Standards, available at http://www.nfpa.org/regs

This edition of *NFPA 70E* was approved as an American National Standard on July 29, 2014.

Foreword to *NFPA 70E*

The Standards Council of the National Fire Protection Association announced on January 7, 1976, the formal appointment of a new electrical standards development committee. Entitled the Committee on Electrical Safety Requirements for Employee Workplaces, *NFPA 70E*, this new committee reported to the Association through the Technical Correlating Committee on National Electrical Code®. This committee was formed to assist OSHA in preparing electrical safety standards that would serve OSHA's needs and that could be expeditiously promulgated through the provisions of Section 6(b) of the Occupational Safety and Health Act. OSHA found that in attempting to utilize the latest edition of *NFPA 70®*, *National Electrical Code® (NEC®)*, it was confronted with the following problems:

(1) Updating to a new edition of the *NEC* would have to be accomplished through the OSHA 6(b) procedures. OSHA adopted the 1968 and then the 1971 *NEC* under Section 6(a) procedures of the Occupational Safety and Health Act of 1970. Today, however, OSHA can only adopt or modify a standard by the procedures of Section 6(b) of the OSHA Act, which provide for public notice, opportunity for public comment, and public hearings. The adoption of a new edition of the *NEC* by these procedures would require extensive effort and application of resources by OSHA and others. Even so, going through the Section 6(b) procedures might result in requirements substantially different from those of the *NEC*, thereby creating the problem of conflict between the OSHA standard and other national and local standards.

(2) The *NEC* is intended for use primarily by those who design, install, and inspect electrical installations. OSHA's electrical regulations address employers and employees in their workplaces. The technical content and complexity of the *NEC* is extremely difficult for the average employer and employee to understand.

(3) Some of the detailed provisions within the *NEC* are not directly related to employee safety and, therefore, are of little value for OSHA's needs.

(4) Requirements for electrical safety-related work practices and maintenance of the electrical system considered critical to safety are not found in the *NEC*, which is essentially an electrical installation document. However, OSHA must also consider and develop these safety areas in its regulations.

With these problem areas, it became apparent that a need existed for a new standard, tailored to fulfill OSHA's responsibilities, that would still be fully consistent with the *NEC*.

The foregoing issues led to the concept that a document be put together by a competent group, representing all interests, that would extract suitable portions from the *NEC* and from other documents applicable to electrical safety. This concept and an offer of assistance was submitted in May 1975 to the Assistant Secretary of Labor for OSHA, who responded as follows: "The concept, procedures, and scope of the effort discussed with my staff for preparing the subject standard appear to have great merit, and an apparent need exists for this proposed consensus document which OSHA could consider for promulgation under the provisions of Section 6(b) of the Act. OSHA does have an interest in this effort and believes the proposed standard would serve a useful purpose." With this positive encouragement from OSHA, a proposal to prepare such a document was presented to the NFPA Electrical Section, which unanimously supported a recommendation that the *NEC* Correlating Committee examine the feasibility of developing a document to be used as a basis for evaluating electrical safety in the workplace. In keeping with the recommendation of the Electrical Section and Correlating Committee, the Standards Council authorized the establishment of a committee to carry out this examination.

The committee found it feasible to develop a standard for electrical installations that would be compatible with the OSHA requirements for safety for the employee in locations covered by the *NEC*. The new standard was visualized as consisting of four major parts: Part I, Installation Safety Requirements; Part II, Safety-Related Work Practices; Part III, Safety-Related Maintenance Requirements; and Part IV, Safety Requirements for Special Equipment. Although desirable, it was not considered essential for all of the parts to be completed before the standard was published and made available. Each part is recognized as being an important aspect of electrical safety in the workplace, but the parts are sufficiently independent of each other to permit their separate publication. The new standard was named *NFPA 70E, Standard for Electrical Safety Requirements for Employee Workplaces*. The first edition was published in 1979 and included only Part I.

The second edition was published in 1981. It included Part I as originally published and a new Part II. In 1983, the third edition included Part I and Part II as originally published and a new Part III. In 1988, the fourth edition was published with only minor revisions.

The fifth edition, published in 1995, included major revisions to Part I, updating it to conform to the 1993 edition of *NFPA 70, National Electrical Code (NEC)*. In Part II of the fifth edition, the concepts of "limits of approach" and establishment of an "arc" were introduced. In 2000, the sixth edition included a complete Part I update to the 1999 *NEC*, as well as a new Part IV. Part II continued to focus on establishing flash protection boundaries and the use of personal protective equipment. Also, added to Part II for 2000 were charts to assist the user in applying appropriate protective clothing and personal protective equipment for common tasks.

The seventh edition, published in 2004, reflected several significant changes to the document. The major changes emphasized safe work practices. Clarity and usability of the document were also enhanced. The name of the document was changed to *NFPA 70E, Standard for Electrical Safety in the Workplace*. The entire document was reformatted to comply with the *National Electrical Code® Style Manual*, providing a unique designation for each requirement. The existing parts were renamed as chapters and were reorganized with the safety-related work practices relocated to the front of the document to highlight the emphasis, followed by safety-related maintenance requirements, safety requirements for special equipment, and safety-related installation requirements. The chapter on safety-related work practices also was reorganized to emphasize working on live parts as the last alternative work practice. An energized electrical work permit and related requirements were incorporated into the document. Several definitions were modified or added to enhance usability of the document, and Chapter 4 was updated to correlate with the 2002 edition of the *NEC*.

Essential to the proper use of Chapter 4 of this standard is the understanding that it is not intended to be applied as a design, an installation, a modification, or a construction standard for an electrical installation or system. Its content was intentionally limited in comparison to the content of the *NEC* in order to apply to an electrical installation or a system as part of an employee's workplace. This standard is compatible with corresponding provisions of the *NEC* but is not intended to be used, nor can it be used, in lieu of the *NEC*.

It can be debated that all of the requirements of the *NEC*, when traced through a chain of events, relate to an electrical hazard, but, for practical purposes, inclusion has not been made of those provisions that, in general, are not directly associated with employee safety. In determining the provisions that should be included in Chapter 4, the following guidelines were used:

(1) The provisions should provide protection to the employee from electrical hazards.

(2) The provisions should be excerpted from the *NEC* in a manner that maintains their intent as they apply to employee safety. In some cases, it has been judged essential to the meaning of the excerpted passages to retain some material not applying to employee safety.

(3) The provisions should be selected in a manner that will reduce the need for frequent revision yet avoid technical obsolescence.

(4) Compliance with the provisions should be determined by means of an inspection during the normal state of employee occupancy without removal of parts requiring shutdown of the electrical installation or damaging the building structure or finish.

(5) The provisions should not be encumbered with unnecessary details.

(6) The provisions should be written to enhance their understanding by the employer and employee.

(7) The provisions must not add any requirements not found in the *NEC*, nor must the intent of the *NEC* be changed if the wording is changed.

Chapter 4 of *NFPA 70E* was, therefore, intended to serve a very specific need of OSHA and is in no way intended to be used as a substitute for the *NEC*. Omission of any requirements presently in the *NEC* does not in any way affect the *NEC*, nor should these omitted requirements be considered as unimportant. They are essential to the *NEC* and its intended application; that is, its use by those who design, install, and inspect electrical installations. *NFPA 70E*, on the other hand, is intended for use by employers, employees, and OSHA.

For 2009, over 1300 proposals and comments were reviewed by the committee, upgrading requirements throughout the document. Among the most significant, Chapter 4 was deleted because it was a duplicate of *National Electrical Code* installation requirements. Since the *NEC* and *NFPA 70E* are on different revision cycles, there was always the risk that the contents of Chapter 4 of *NFPA 70E* were not up to date with the *NEC*. Article 350 was added for R&D facilities. Other changes included significant revisions to Annex D, Annex F, and Annex J and the addition of Annex M, Annex N, and Annex O.

The 2012 edition of *NFPA 70E* marked another waypoint as this standard continued to evolve and meet the electrical safety needs of employers and employees. New research, new technology, and technical input from users of the standard provided the foundation for new and revised requirements that addressed the electrical hazards encountered by employees in today's workplaces. Revisions that expanded or clarified requirements in the 2009 edition, inclusion of new technical material that had not been covered by previous editions of the standard, and removal of requirements that were related to the safe installation of electrical equipment (particularly from Article 320) rather than being safe electrical work practices were some of the major actions undertaken by the Technical Committee on Electrical Safety in the Workplace in the 2012 revision cycle. In addition, provisions throughout the standard covering the separate but directly related concepts of hazard identification and risk assessment were revised to clarify these concepts. A significant revision to Annex F provided extensive coverage of this topic to assist users of the standard with implementing effective hazard identification and risk assessment procedures.

In the 2012 revision cycle, the majority of changes occurred in Chapter 1. With the exception of the major revisions in Article 320, Safety Requirements Related to Batteries and Battery Rooms, the revisions in Chapters 2 and 3 were primarily for clarification and editorial purposes. In addition to Annex F, Annexes D, H, J, and O saw substantive revisions. Annex P on aligning *NFPA 70E* implementation with occupational health and safety management standards was added. Some of the major revisions included changing "flame-resistant (FR)" to "arc-rated (AR)" in regard to personal protective equipment (PPE) throughout the standard. Article 105, Application of Safety-Related Work Practices, and a requirement for hearing protection when working within arc flash boundary were added, as were work practice requirements on the use of GFCIs to protect employees. Clarification was provided that Article 130 applies whether incident energy analysis or the hazard/risk table was used to determine use and level of PPE. Short-circuit current, fault clearing time, and potential arc flash boundary information were included to each of the major categories in the hazard/risk category tables.

The 2015 edition of *NFPA 70E* reflects a major shift in how stakeholders evaluate electrical risk. In support of this, new definitions were added to Article 100, including *Hazard, Hazardous, Risk*, and *Risk Assessment*. Throughout the document changes were made to provide clarity to users, such as changing "arc flash hazard analysis" to "arc flash risk assessment," "shock hazard analysis" to "shock risk assessment," "electrical hazard analysis" to "electrical hazard risk assessment," and "hazard identification and risk assessment" to "risk assessment." These global changes ensure consistent use of these terms throughout the document and provide consistency between *NFPA 70E* and other standards that address hazards and risk. Other major revisions include the following:

(1) Safety-related maintenance requirements and other administrative controls were added to the Scope statement of the standard to clarify that training and auditing are equally important safety-related work practices. [90.2(A)]

(2) The definition of *Bare-Hand Work* and all references to bare-hand work were removed. The term, which is considered to be a "utility type" line work technique, is more appropriately addressed in other standards.

(3) The definition for *Qualified Person* was revised to correlate the definition with OSHA 1910.399 Note 2. [100]

(4) *Prohibited Approach Boundary* was deleted. The requirement for using shock protective equipment typically begins at the restricted approach boundary. No additional protective equipment was required when crossing the prohibited approach boundary. Previous changes that used the limited approach boundary or arc flash boundary for "triggering" requirements have made the use of this term unnecessary.

(5) An electrical safety program must now include elements that consider condition of maintenance. [110.1(B)]

(6) Audits of field work to verify compliance with the procedures of the electrical safety program must be performed at intervals not to exceed 1 year. [110.1(I)(2)]

(7) The location, sizing, and application of temporary protective grounding equipment is required to be identified as part of the employer's job planning. [120.3(A)]

(8) New requirements clarifying where normal operation of electric equipment is permitted were added. The equipment must be properly installed and maintained, equipment doors closed and secured, and all covers in place and secured, and there is no evidence of impending failure. [130.2(A)(4)]

(9) Clarification was added that either the incident energy analysis method or arc flash PPE categories method can be used on the same piece of equipment for the selection of PPE, but not both. The revision also clarifies that the results of an incident energy analysis to specify an arc flash PPE category in Table 130.7(C)(16) is not permitted. [130.5(C)]

(10) Field-marked equipment labeling requirements were revised to require the label to be updated where the arc flash hazard risk assessment identifies a change that renders the label inaccurate. Also, the documentation, installation, and maintenance of the field-marked label is the responsibility of the owner of the electrical equipment. [130.5(D)]

(11) Additional text now provides the user with a boundary to the existing requirements in 130.6(D). Conductive articles being worn shall not be worn within the restricted approach boundary or where they present an electrical contact hazard. [130.6(D)]

(12) A new task-based table combines the separate ac and dc tables previously used to determine when arc flash PPE is required and makes them consistent, improving usability. The new table lists the task, equipment condition, and arc flash PPE required. It utilizes a simple yes or no format if arc flash PPE is required. [130.7(C)(15)(A)(a)]

(13) New equipment-based tables were added for determining the arc flash PPE category, Table 130.7(C)(15)(A)(b) for ac systems and Table 130.7(C)(15)(B) for dc systems. The tables list the equipment, arc flash PPE category, and the arc flash boundary. [Table 130.7(C)(15)(A)(b), Table 130.7(C)(15)(B)]

(14) Hazard/risk category 0 has been removed from Table 130.7(C)(16). Hazard/risk category will now be referred to as PPE category. Hazard/risk category 0 was deleted because the new PPE table only specifies PPE for work within the arc flash boundary. If there is no arc flash hazard, then no arc flash PPE is required and it is therefore not necessary on a table devoted to PPE. [Table 130.7(C)(16)]

(15) The criterion for employees to use insulated tools or handling equipment has been changed from the limited approach boundary to restricted approach boundary. [130.7(D)(1)]

(16) Barricades cannot be placed closer than the limited approach boundary. Where the arc flash boundary is greater than the limited approach boundary, barricades cannot be placed closer than the arc flash boundary. [130.7(E)(2)]

(17) A new section is added requiring the employer to perform a risk assessment before cutting or drilling into equipment, floors, walls, or structural elements where a likelihood of contacting energized electrical lines or parts exists. [130.10]

(18) Clarification is provided that the equipment owner or the owner's designated representative is responsible for maintenance of the electrical equipment and documentation. [205.3]

(19) New maintenance requirements for test instruments and associated test leads utilized in the verification of the absence or presence of voltages were added. The maintenance program for test instruments must include functional verification as described in 110.4(A)(5). [250.4]

(20) New section 320.3(A)(1) requires a risk assessment to be performed prior to any work on a battery system to identify the chemical, electrical shock, and arc flash hazards and assess the risks associated with the type of tasks to be performed. [320.3(A)(1)]

Contents

Correlating Committee on National Electrical Code®

Michael J. Johnston, *Chair*
National Electrical Contractors Association, MD [IM]

Mark W. Earley, *Secretary (Nonvoting)*
National Fire Protection Association, MA

Kimberly L. Shea, *Recording Secretary (Nonvoting)*
National Fire Protection Association, MA

James E. Brunssen, Telcordia, NJ [UT]
Rep. Alliance for Telecommunications Industry Solutions
Merton W. Bunker, Jr., U.S. Department of State, VA [U]
William R. Drake, Actuant Electrical, CA [M]
William T. Fiske, Intertek Testing Services, NY [RT]
Palmer L. Hickman, Electrical Training Alliance, MD [L]
Rep. International Brotherhood of Electrical Workers
David L. Hittinger, Independent Electrical Contractors of Greater Cincinnati, OH [IM]
Rep. Independent Electrical Contractors, Inc.
Daniel J. Kissane, Legrand/Pass & Seymour, NY [M]
Rep. National Electrical Manufacturers Association
John R. Kovacik, UL LLC, IL [RT]
Neil F. LaBrake, Jr., National Grid, NY [UT]
Rep. Electric Light & Power Group/EEI
Danny Liggett, The DuPont Company, Inc., TX [U]
Rep. American Chemistry Council
Richard P. Owen, Oakdale, MN [E]
Rep. International Association of Electrical Inspectors

Alternates

Thomas L. Adams, Engineering Consultant, IL [UT]
(Alt. to N. F. LaBrake, Jr.)
Lawrence S. Ayer, Biz Com Electric, Inc., OH [IM]
(Alt. to D. L. Hittinger)
James T. Dollard, Jr., IBEW Local Union 98, PA [L]
(Alt. to P. L. Hickman)
Stanley J. Folz, Morse Electric Company, NV [IM]
(Alt. to M. J. Johnston)
Ernest J. Gallo, Telcordia Technologies (Ericsson), NJ [UT]
(Alt. to J. E. Brunssen)
Alan Manche, Schneider Electric, KY [M]
(Alt. to W. R. Drake)
Robert A. McCullough, Tuckerton, NJ [E]
(Alt. to R. P. Owen)
Michael E. McNeil, FMC Bio Polymer, ME [U]
(Alt. to D. Liggett)
Mark C. Ode, UL LLC, AZ [RT]
(Alt. to J. R. Kovacik)
James F. Pierce, Intertek, OR [RT]
(Alt. to W. T. Fiske)
Vincent J. Saporita, Eaton's Bussmann Business, MO [M]
(Alt. to D. J. Kissane)

Nonvoting

Timothy J. Pope, Canadian Standards Association, Canada [SE]
Rep. CSA/Canadian Electrical Code Committee
D. Harold Ware, Libra Electric Company, OK [IM]
(Member Emeritus)

Committee Scope: This Committee shall have primary responsibility for documents on minimizing the risk of electricity as a source of electric shock and as a potential ignition source of fires and explosions. It shall also be responsible for text to minimize the propagation of fire and explosions due to electrical installations.

Technical Committee on Electrical Safety in the Workplace

David A. Dini, *Chair*
UL LLC, IL [RT]

Louis A. Barrios, Shell Global Solutions, TX [U]
Rep. American Petroleum Institute
William Bruce Bowman, Fox Systems, Inc., GA [IM]
Rep. Independent Electrical Contractors, Inc.
Steven C. Chybowski, Rockwell Automation Inc., WI [M]
Carey J. Cook, S&C Electric Company, IL [M]
Michael J. Douglas, General Motors Corporation, MI [U]
Drake A. Drobnick, Saline, MI [SE]
Marcia L. Eblen, Pacific Gas & Electric, CA [SE]
Rep. ASTM F18-Electrical Protective Equipment for Workers
Bobby J. Gray, Hoydar/Buck, Inc., WA [E]
Lee R. Hale, Lee Hale Consulting, Inc., IA [M]
Rep. The Aluminum Association, Inc.
James B. Hayes, Florida Institute of Technology, FL [U]
Palmer L. Hickman, Electrical Training Alliance, MD [L]
Rep. International Brotherhood of Electrical Workers
John Luke, The ESCO Group, IA [IM]
Rep. National Electrical Contractors Association
Michael J. Madrigal, Kansas City Power & Light, KS [U]
Rep. Edison Electric Institute
Mark McNellis, Sandia National Laboratories, NM [U]
Daleep C. Mohla, DCM Electrical Consulting Services, Inc., TX [SE]
Rep. Institute of Electrical & Electronics Engineers, Inc.
Dennis K. Neitzel, AVO Training Institute, Inc., TX [SE]
David A. Pace, Olin Corporation, AL [U]
Rep. American Chemistry Council
Vincent J. Saporita, Eaton's Bussman Business, MO [M]
Rep. National Electrical Manufacturers Association
James G. Stallcup, Grayboy, Inc., TX [SE]
Charlie R. Thurmond III, ThyssenKrupp Elevator, TN [IM]
John M. Tobias, U.S. Department of the Army, MD [U]
David M. Wallis, U.S. Department of Labor, DC [E]
Rodney J. West, Schneider Electric, OH [M]
Ron Widup, Shermco Industries, TX [IM]
Rep. InterNational Electrical Testing Association

Alternates

Lawrence S. Ayer, Biz Com Electric, Inc., OH [IM]
(Alt. to W. B. Bowman)
John F. Boothroyd, Entergy Fossil Operations, TX [U]
(Alt. to M. J. Madrigal)
Steven D. Corrado, UL LLC, NC [RT]
(Alt. to D. A. Dini)
Daryld Ray Crow, DRC Consulting, Ltd., WA [M]
(Alt. to L. R. Hale)
Paul Dobrowsky, Innovative Technology Services, NY [SE]
(Alt. to D. C. Mohla)
James T. Dollard, Jr., IBEW Local Union 98, PA [L]
(Alt. to P. L. Hickman)
Heath Garrison, National Renewable Energy Laboratory, CO [U]
(Alt. to M. McNellis)
Eric Glaude, Chevron, TX [U]
(Alt. to L. A. Barrios)
Danny Liggett, The DuPont Company, Inc., TX [U]
(Alt. to D. A. Pace)
Kevin J. Lippert, Eaton Corporation, PA [M]
(Alt. to V. J. Saporita)
Charles R. Miller, Lighthouse Educational Services, TN [SE]
(Alt. to D. A. Drobnick)
Thomas D. Norwood, AVO Training Institute, TX [SE]
(Alt. to D. K. Neitzel)
Larry D. Perkins, US Department of Energy, TN [E]
(Alt. to B. J. Gray)
James W. Stallcup, Jr., Grayboy, Inc., TX [SE]
(Alt. to J. G. Stallcup)
Samuel B. Stonerock, Southern California Edison Company, CA [SE]
(Alt. to M. L. Eblen)
Terry Tiede, Schneider Electric, IA [M]
(Alt. to R. J. West)
James R. White, Shermco Industries, Inc., TX [IM]
(Alt. to R. Widup)

Nonvoting

Mike Doherty, Powertel Utilities Contractors Limited, Canada [RT]
Rep. Canadian Standards Association
Kerry Heid, Magna Electric Corporation, Canada [U]
Rep. Canadian Standards Association
James R. Tomaseski, International Brotherhood of Electrical Workers, DC [L]
Rep. National Electrical Safety Code
Michael Fontaine, NFPA Staff Liaison

Committee Scope: This Committee shall have primary responsibility for documents for work practices that are necessary to provide a practical safe workplace relative to the hazards associated with electrical energy. This Committee shall have primary jurisdiction, but shall report to Technical Correlating Committee of the National Electrical Code.

NFPA 70E®

Standard for Electrical Safety in the Workplace®

2015 Edition

IMPORTANT NOTE: This NFPA document is made available for use subject to important notices and legal disclaimers. These notices and disclaimers appear in all publications containing this document and may be found under the heading "Important Notices and Disclaimers Concerning NFPA Documents." They can also be obtained on request from NFPA or viewed at www.nfpa.org/disclaimers.

This 2015 edition includes the following usability features as aids to the user. Changes other than editorial are highlighted with gray shading within sections and with vertical ruling for large blocks of changed or new text and for new tables and changed or new figures. Where one or more complete paragraphs have been deleted, the deletion is indicated by a bullet (•) between the paragraphs that remain. The index now has dictionary-style headers with helpful identifiers at the top of every index page.

A reference in brackets [ ] following a section or paragraph indicates material that has been extracted from another NFPA document. As an aid to the user, the complete title and edition of the source documents for extracts are given in Annex A. Extracted text may be edited for consistency and style and may include the revision of internal paragraph references and other references as appropriate. Requests for interpretations or revisions of extracted text shall be sent to the technical committee responsible for the source document.

Information on referenced publications can be found in Informative Annex A and Informative Annex B.

ARTICLE 90
Introduction

90.1 Purpose. The purpose of this standard is to provide a practical safe working area for employees relative to the hazards arising from the use of electricity.

90.2 Scope.

(A) Covered. This standard addresses electrical safety-related work practices, safety-related maintenance requirements, and other administrative controls for employee workplaces that are necessary for the practical safeguarding of employees relative to the hazards associated with electrical energy during activities such as the installation, inspection, operation, maintenance, and demolition of electric conductors, electric equipment, signaling and communications conductors and equipment, and raceways. This standard also includes safe work practices for employees performing other work activities that can expose them to electrical hazards as well as safe work practices for the following:

(1) Installation of conductors and equipment that connect to the supply of electricity
(2) Installations used by the electric utility, such as office buildings, warehouses, garages, machine shops, and recreational buildings that are not an integral part of a generating plant, substation, or control center

Informational Note: This standard addresses safety of workers whose job responsibilities entail interaction with electrical equipment and systems with potential exposure to energized electrical equipment and circuit parts. Concepts in this standard are often adapted to other workers whose exposure to electrical hazards is unintentional or not recognized as part of their job responsibilities. The highest risk for injury from electrical hazards for other workers involve unintentional contact with overhead power lines and electric shock from machines, tools, and appliances.

(B) Not Covered. This standard does not cover safety-related work practices for the following:

(1) Installations in ships, watercraft other than floating buildings, railway rolling stock, aircraft, or automotive vehicles other than mobile homes and recreational vehicles

•

(2) Installations of railways for generation, transformation, transmission, or distribution of power used exclusively for operation of rolling stock or installations used exclusively for signaling and communications purposes
(3) Installations of communications equipment under the exclusive control of communications utilities located outdoors or in building spaces used exclusively for such installations
(4) Installations under the exclusive control of an electric utility where such installations:
   a. Consist of service drops or service laterals, and associated metering, or
   b. Are located in legally established easements or rights-of-way designated by or recognized by public service commissions, utility commissions, or other regulatory agencies having jurisdiction for such installations, or
   c. Are on property owned or leased by the electric utility for the purpose of communications, metering, generation, control, transformation, transmission, or distribution of electric energy, or
   d. Are located by other written agreements either designated by or recognized by public service commissions, utility commissions, or other regulatory agencies having jurisdiction for such installations. These written agreements shall be limited to installations

for the purpose of communications, metering, generation, control, transformation, transmission, or distribution of electric energy where legally established easements or rights-of-way cannot be obtained. These installations shall be limited to federal lands, Native American reservations through the U.S. Department of the Interior Bureau of Indian Affairs, military bases, lands controlled by port authorities and state agencies and departments, and lands owned by railroads.

90.3 Standard Arrangement. This standard is divided into the introduction and three chapters, as shown in Figure 90.3. Chapter 1 applies generally for safety-related work practices; Chapter 2 applies to safety-related maintenance requirements for electrical equipment and installations in workplaces; and Chapter 3 supplements or modifies Chapter 1 with safety requirements for special equipment.

• Informative annexes are not part of the requirements of this standard but are included for informational purposes only.

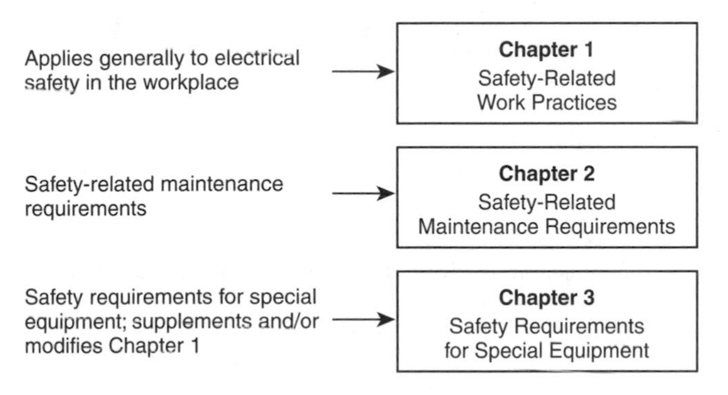

Figure 90.3 Standard Arrangement.

90.4 Organization. This standard is divided into the following 3 chapters and 16 informative annexes:

(1) Chapter 1, Safety-Related Work Practices
(2) Chapter 2, Safety-Related Maintenance Requirements
(3) Chapter 3, Safety Requirements for Special Equipment
(4) Informative Annex A, Referenced Publications
(5) Informative Annex B, Informational References
(6) Informative Annex C, Limits of Approach
(7) Informative Annex D, Incident Energy and Arc Flash Boundary Calculation Methods
(8) Informative Annex E, Electrical Safety Program
(9) Informative Annex F, Risk Assessment Procedure
(10) Informative Annex G, Sample Lockout/Tagout Procedure
(11) Informative Annex H, Guidance on Selection of Protective Clothing and Other Personal Protective Equipment
(12) Informative Annex I, Job Briefing and Planning Checklist
(13) Informative Annex J, Energized Electrical Work Permit
(14) Informative Annex K, General Categories of Electrical Hazards
(15) Informative Annex L, Typical Application of Safeguards in the Cell Line Working Zone
(16) Informative Annex M, Layering of Protective Clothing and Total System Arc Rating
(17) Informative Annex N, Example Industrial Procedures and Policies for Working Near Overhead Electrical Lines and Equipment
(18) Informative Annex O, Safety-Related Design Requirements
(19) Informative Annex P, Aligning Implementation of This Standard with Occupational Health and Safety Management Standards

90.5 Mandatory Rules, Permissive Rules, and Explanatory Material.

(A) Mandatory Rules. Mandatory rules of this standard are those that identify actions that are specifically required or prohibited and are characterized by the use of the terms *shall* or *shall not.*

(B) Permissive Rules. Permissive rules of this standard are those that identify actions that are allowed but not required, are normally used to describe options or alternative methods, and are characterized by the use of the terms *shall be permitted* or *shall not be required.*

(C) Explanatory Material. Explanatory material, such as references to other standards, references to related sections of this standard, or information related to a rule in this standard, is included in this standard in the form of informational notes. Such notes are informational only and are not enforceable as requirements of this standard.

Brackets containing section references to another NFPA document are for informational purposes only and are provided as a guide to indicate the source of the extracted text. These bracketed references immediately follow the extracted text.

Informational Note: The format and language used in this standard follow guidelines established by NFPA and published in the *National Electrical Code Style Manual*. Copies of this manual can be obtained from NFPA.

90.6 Formal Interpretations. To promote uniformity of interpretation and application of the provisions of this standard, formal interpretation procedures have been established and are found in the NFPA Regulations Governing Committee Projects.

Chapter 1 Safety-Related Work Practices

ARTICLE 100 Definitions

Scope. This article contains only those definitions essential to the proper application of this standard. It is not intended to include commonly defined general terms or commonly defined technical terms from related codes and standards. In general, only those terms that are used in two or more articles are defined in Article 100. Other definitions are included in the article in which they are used but may be referenced in Article 100. The definitions in this article shall apply wherever the terms are used throughout this standard.

Accessible (as applied to equipment). Admitting close approach; not guarded by locked doors, elevation, or other effective means. [**70:**100]

Accessible (as applied to wiring methods). Capable of being removed or exposed without damaging the building structure or finish or not permanently closed in by the structure or finish of the building. [**70:**100]

Accessible, Readily (Readily Accessible). Capable of being reached quickly for operation, renewal, or inspections without requiring those to whom ready access is requisite to actions such as to use tools, to climb over or remove obstacles, or to resort to portable ladders, and so forth. [**70:**100]

Approved. Acceptable to the authority having jurisdiction.

Arc Flash Hazard. A dangerous condition associated with the possible release of energy caused by an electric arc.

Informational Note No. 1: An arc flash hazard may exist when energized electrical conductors or circuit parts are exposed or when they are within equipment in a guarded or enclosed condition, provided a person is interacting with the equipment in such a manner that could cause an electric arc. Under normal operating conditions, enclosed energized equipment that has been properly installed and maintained is not likely to pose an arc flash hazard.

Informational Note No. 2: See Table 130.7(C)(15)(A)(a) for examples of activities that could pose an arc flash hazard.

Arc Flash Suit. A complete arc-rated clothing and equipment system that covers the entire body, except for the hands and feet.

Informational Note: An arc flash suit may include pants or overalls, a jacket or a coverall, and a beekeeper-type hood fitted with a face shield.

Arc Rating. The value attributed to materials that describes their performance to exposure to an electrical arc discharge. The arc rating is expressed in cal/cm^2 and is derived from the determined value of the arc thermal performance value (ATPV) or energy of breakopen threshold (E_{BT}) (should a material system exhibit a breakopen response below the ATPV value). Arc rating is reported as either ATPV or E_{BT}, whichever is the lower value.

Informational Note No. 1: Arc-rated clothing or equipment indicates that it has been tested for exposure to an electric arc. Flame resistant clothing without an arc rating has not been tested for exposure to an electric arc. All arc-rated clothing is also flame-resistant.

Informational Note No. 2: *Breakopen* is a material response evidenced by the formation of one or more holes in the innermost layer of arc-rated material that would allow flame to pass through the material.

Informational Note No. 3: ATPV is defined in ASTM F1959/F1959M, *Standard Test Method for Determining the Arc Rating of Materials for Clothing*, as the incident energy (cal/cm^2) on a material or a multilayer system of materials that results in a 50 percent probability that sufficient heat transfer through the tested specimen is predicted to cause the onset of a second degree skin burn injury based on the Stoll curve.

Informational Note No. 4: E_{BT} is defined in ASTM F1959/F1959M, *Standard Test Method for Determining the Arc Rating of Materials for Clothing*, as the incident energy (cal/cm^2) on a material or a material system that results in a 50 percent probability of breakopen. Breakopen is defined as a hole with an area of 1.6 cm^2 (0.5 in^2) or an opening of 2.5 cm (1.0 in.) in any dimension.

Attachment Plug (Plug Cap) (Plug). A device that, by insertion in a receptacle, establishes a connection between the conductors of the attached flexible cord and the conductors connected permanently to the receptacle. [**70:**100]

Authority Having Jurisdiction (AHJ). An organization, office, or individual responsible for enforcing the requirements of a code or standard, or for approving equipment, materials, an installation, or a procedure.

Informational Note: The phrase "authority having jurisdiction," or its acronym AHJ, is used in NFPA documents in a broad manner, since jurisdictions and approval agencies vary, as do their responsibilities. Where public safety is primary, the authority having jurisdiction may be a federal, state, local, or other regional department or individual such as a fire chief; fire marshal; chief of a fire prevention bureau, labor department, or health department; building official; electrical inspector; or others having statutory authority. For insurance purposes, an insurance inspection department, rating bureau, or other insurance company representative may be the authority having jurisdiction. In many circumstances, the property owner or his or her designated agent assumes the role of the authority having jurisdiction; at government installations, the commanding officer or departmental official may be the authority having jurisdiction.

Automatic. Performing a function without the necessity of human intervention.

Balaclava (Sock Hood). An arc-rated hood that protects the neck and head except for the facial area of the eyes and nose.

• **Barricade.** A physical obstruction such as tapes, cones, or A-frame-type wood or metal structures intended to provide a warning and to limit access.

Barrier. A physical obstruction that is intended to prevent contact with equipment or energized electrical conductors and circuit parts or to prevent unauthorized access to a work area.

Bonded (Bonding). Connected to establish electrical continuity and conductivity. [**70:**100]

Bonding Conductor or Jumper. A reliable conductor to ensure the required electrical conductivity between metal parts required to be electrically connected. [**70:**100]

Boundary, Arc Flash. When an arc flash hazard exists, an approach limit at a distance from a prospective arc source within which a person could receive a second degree burn if an electrical arc flash were to occur.

Informational Note: A second degree burn is possible by an exposure of unprotected skin to an electric arc flash above the incident energy level of 5 J/cm^2 (1.2 cal/cm^2).

Boundary, Limited Approach. An approach limit at a distance from an exposed energized electrical conductor or circuit part within which a shock hazard exists.

• **Boundary, Restricted Approach.** An approach limit at a distance from an exposed energized electrical conductor or circuit part within which there is an increased likelihood of electric shock, due to electrical arc-over combined with inadvertent movement, for personnel working in close proximity to the energized electrical conductor or circuit part.

Branch Circuit. The circuit conductors between the final overcurrent device protecting the circuit and the outlet(s). [**70:**100]

Building. A structure that stands alone or that is cut off from adjoining structures by fire walls with all openings therein protected by approved fire doors. [**70:**100]

Cabinet. An enclosure that is designed for either surface mounting or flush mounting and is provided with a frame, mat, or trim in which a swinging door or doors are or can be hung. [**70:**100]

Circuit Breaker. A device designed to open and close a circuit by nonautomatic means and to open the circuit automatically on a predetermined overcurrent without damage to itself when properly applied within its rating. [**70:**100]

Informational Note: The automatic opening means can be integral, direct acting with the circuit breaker, or remote from the circuit breaker.

Conductive. Suitable for carrying electric current.

Conductor, Bare. A conductor having no covering or electrical insulation whatsoever. [**70:**100]

Conductor, Covered. A conductor encased within material of composition or thickness that is not recognized by this *Code* as electrical insulation. [**70:**100]

Conductor, Insulated. A conductor encased within material of composition and thickness that is recognized by this *Code* as electrical insulation. [**70:**100]

Controller. A device or group of devices that serves to govern, in some predetermined manner, the electric power delivered to the apparatus to which it is connected. [**70:**100]

Current-Limiting Overcurrent Protective Device. A device that, when interrupting currents in its current-limiting range, reduces the current flowing in the faulted circuit to a magnitude substantially less than that obtainable in the same circuit if the device were replaced with a solid conductor having comparable impedance.

Cutout. An assembly of a fuse support with either a fuseholder, fuse carrier, or disconnecting blade. The fuseholder or fuse carrier may include a conducting element (fuse link), or may act as the disconnecting blade by the inclusion of a nonfusible member.

De-energized. Free from any electrical connection to a source of potential difference and from electrical charge; not having a potential different from that of the earth.

Device. A unit of an electrical system, other than a conductor, that carries or controls electric energy as its principal function. [**70:**100]

Disconnecting Means. A device, or group of devices, or other means by which the conductors of a circuit can be disconnected from their source of supply. [**70:**100]

Disconnecting (or Isolating) Switch (Disconnector, Isolator). A mechanical switching device used for isolating a circuit or equipment from a source of power.

Dwelling Unit. A single unit providing complete and independent living facilities for one or more persons, including permanent provisions for living, sleeping, cooking, and sanitation. [**70:**100]

Electrical Hazard. A dangerous condition such that contact or equipment failure can result in electric shock, arc flash burn, thermal burn, or blast.

Informational Note: Class 2 power supplies, listed low voltage lighting systems, and similar sources are examples of circuits or systems that are not considered an electrical hazard.

Electrical Safety. Recognizing hazards associated with the use of electrical energy and taking precautions so that hazards do not cause injury or death.

Electrically Safe Work Condition. A state in which an electrical conductor or circuit part has been disconnected from energized parts, locked/tagged in accordance with established standards, tested to ensure the absence of voltage, and grounded if determined necessary.

Enclosed. Surrounded by a case, housing, fence, or wall(s) that prevents persons from accidentally contacting energized parts. [**70:**100]

Enclosure. The case or housing of apparatus — or the fence or walls surrounding an installation to prevent personnel from accidentally contacting energized electrical conductors or circuit parts or to protect the equipment from physical damage.

Energized. Electrically connected to, or is, a source of voltage. [**70:**100]

Equipment. A general term, including fittings, devices, appliances, luminaires, apparatus, machinery, and the like, used as a part of, or in connection with, an electrical installation. [**70:**100]

Exposed (as applied to energized electrical conductors or circuit parts). Capable of being inadvertently touched or approached nearer than a safe distance by a person. It is applied to electrical conductors or circuit parts that are not suitably guarded, isolated, or insulated.

Exposed (as applied to wiring methods). On or attached to the surface or behind panels designed to allow access. [**70:**100]

Fitting. An accessory such as a locknut, bushing, or other part of a wiring system that is intended primarily to perform a mechanical rather than an electrical function. [**70:**100]

Fuse. An overcurrent protective device with a circuit-opening fusible part that is heated and severed by the passage of overcurrent through it.

Informational Note: A fuse comprises all the parts that form a unit capable of performing the prescribed functions. It may or may not be the complete device necessary to connect it into an electrical circuit.

Ground. The earth. [**70:**100]

Ground Fault. An unintentional, electrically conducting connection between an ungrounded conductor of an electrical circuit and the normally non–current-carrying conductors, metallic enclosures, metallic raceways, metallic equipment, or earth.

Grounded (Grounding). Connected (connecting) to ground or to a conductive body that extends the ground connection. [**70:**100]

Grounded, Solidly. Connected to ground without inserting any resistor or impedance device. [**70:**100]

Grounded Conductor. A system or circuit conductor that is intentionally grounded. [**70:**100]

Ground-Fault Circuit Interrupter (GFCI). A device intended for the protection of personnel that functions to de-energize a circuit or portion thereof within an established period of time when a current to ground exceeds the values established for a Class A device. [**70:**100]

Informational Note: Class A ground-fault circuit-interrupters trip when the current to ground is 6 mA or higher and do not trip when the current to ground is less than 4 mA. For further information, see ANSI/UL 943, *Standard for Ground-Fault Circuit Interrupters.*

Grounding Conductor, Equipment (EGC). The conductive path(s) that provides a ground-fault current path and connects normally non–current-carrying metal parts of equipment together and to the system grounded conductor or to the grounding electrode conductor, or both. [**70:**100]

Informational Note No. 1: It is recognized that the equipment grounding conductor also performs bonding.

Informational Note No. 2: See 250.118 of *NFPA 70*, *National Electrical Code*, for a list of acceptable equipment grounding conductors.

Grounding Electrode. A conducting object through which a direct connection to earth is established. [**70:**100]

Grounding Electrode Conductor. A conductor used to connect the system grounded conductor or the equipment to a grounding electrode or to a point on the grounding electrode system. [**70:**100]

Guarded. Covered, shielded, fenced, enclosed, or otherwise protected by means of suitable covers, casings, barriers, rails, screens, mats, or platforms to remove the likelihood of approach or contact by persons or objects to a point of danger. [**70:**100]

Hazard. A source of possible injury or damage to health.

Hazardous. Involving exposure to at least one hazard.

Incident Energy. The amount of thermal energy impressed on a surface, a certain distance from the source, generated during an electrical arc event. Incident energy is typically expressed in calories per square centimeter (cal/cm^2).

Incident Energy Analysis. A component of an arc flash risk assessment used to predict the incident energy of an arc flash for a specified set of conditions.

Insulated. Separated from other conducting surfaces by a dielectric (including air space) offering a high resistance to the passage of current.

Informational Note: When an object is said to be insulated, it is understood to be insulated for the conditions to which it is normally subject. Otherwise, it is, within the purpose of these rules, uninsulated.

Interrupter Switch. A switch capable of making, carrying, and interrupting specified currents.

Interrupting Rating. The highest current at rated voltage that a device is identified to interrupt under standard test conditions. [**70:**100]

Informational Note: Equipment intended to interrupt current at other than fault levels may have its interrupting rating implied in other ratings, such as horsepower or locked rotor current.

Isolated (as applied to location). Not readily accessible to persons unless special means for access are used. [**70:**100]

Labeled. Equipment or materials to which has been attached a label, symbol, or other identifying mark of an organization that is acceptable to the authority having jurisdiction and concerned with product evaluation, that maintains periodic inspection of production of labeled equipment or materials, and by whose labeling the manufacturer indicates compliance with appropriate standards or performance in a specified manner.

Listed. Equipment, materials, or services included in a list published by an organization that is acceptable to the authority having jurisdiction and concerned with evaluation of products or services, that maintains periodic inspection of production of listed equipment or materials or periodic evaluation of services, and whose listing states that either the equipment, material, or service meets appropriate designated standards or has been tested and found suitable for a specified purpose.

Informational Note: The means for identifying listed equipment may vary for each organization concerned with product evaluation; some organizations do not recognize equipment as listed unless it is also labeled. The authority having jurisdiction should utilize the system employed by the listing organization to identify a listed product.

Luminaire. A complete lighting unit consisting of a light source, such as a lamp or lamps, together with the parts designed to position the light source and connect it to the power supply. It may also include parts to protect the light source or the ballast or to distribute the light. A lampholder itself is not a luminaire. [**70:**100]

Motor Control Center. An assembly of one or more enclosed sections having a common power bus and principally containing motor control units. [**70:**100]

Outlet. A point on the wiring system at which current is taken to supply utilization equipment. [**70:**100]

Overcurrent. Any current in excess of the rated current of equipment or the ampacity of a conductor. It may result from overload, short circuit, or ground fault. [**70:**100]

Informational Note: A current in excess of rating may be accommodated by certain equipment and conductors for a given set of conditions. Therefore, the rules for overcurrent protection are specific for particular situations.

Overload. Operation of equipment in excess of normal, full-load rating, or of a conductor in excess of rated ampacity that, when it persists for a sufficient length of time, would cause damage or dangerous overheating. A fault, such as a short circuit or ground fault, is not an overload. [**70:**100]

Panelboard. A single panel or group of panel units designed for assembly in the form of a single panel, including buses and automatic overcurrent devices, and equipped with or without switches for the control of light, heat, or power circuits; designed to be placed in a cabinet or cutout box placed in or against a wall, partition, or other support; and accessible only from the front. [**70:**100]

Premises Wiring (System). Interior and exterior wiring, including power, lighting, control, and signal circuit wiring together with all their associated hardware, fittings, and wiring devices, both permanently and temporarily installed. This includes: (a) wiring from the service point or power source to the outlets; or (b) wiring from and including the power source to the outlets where there is no service point.

Such wiring does not include wiring internal to appliances, luminaires, motors, controllers, motor control centers, and similar equipment. [**70:**100]

Informational Note: Power sources include, but are not limited to, interconnected or stand-alone batteries, solar photovoltaic systems, other distributed generation systems, or generators.

Qualified Person. One who has demonstrated skills and knowledge related to the construction and operation of electrical equipment and installations and has received safety training to identify and avoid the hazards involved.

Raceway. An enclosed channel of metal or nonmetallic materials designed expressly for holding wires, cables, or busbars, with additional functions as permitted in this standard. [**70:**100]

Receptacle. A receptacle is a contact device installed at the outlet for the connection of an attachment plug. A single receptacle is a single contact device with no other contact device on the same yoke. A multiple receptacle is two or more contact devices on the same yoke. [**70:**100]

Risk. A combination of the likelihood of occurrence of injury or damage to health and the severity of injury or damage to health that results from a hazard.

Risk Assessment. An overall process that identifies hazards, estimates the potential severity of injury or damage to health, estimates the likelihood of occurrence of injury or damage to health, and determines if protective measures are required.

Informational Note: As used in this standard, *arc flash risk assessment* and *shock risk assessment* are types of risk assessments.

Service Drop. The overhead conductors between the utility electric supply system and the service point. [**70:**100]

Service Lateral. The underground conductors between the utility electric supply system and the service point. [**70:**100]

Service Point. The point of connection between the facilities of the serving utility and the premises wiring. [**70:**100]

Informational Note: The service point can be described as the point of demarcation between where the serving utility ends and the premises wiring begins. The serving utility generally specifies the location of the service point based on the conditions of service.

Shock Hazard. A dangerous condition associated with the possible release of energy caused by contact or approach to energized electrical conductors or circuit parts.

Short-Circuit Current Rating. The prospective symmetrical fault current at a nominal voltage to which an apparatus or system is able to be connected without sustaining damage exceeding defined acceptance criteria. [**70:**100]

Single-Line Diagram. A diagram that shows, by means of single lines and graphic symbols, the course of an electric circuit or system of circuits and the component devices or parts used in the circuit or system.

Special Permission. The written consent of the authority having jurisdiction. [**70:**100]

Step Potential. A ground potential gradient difference that can cause current flow from foot to foot through the body.

Structure. That which is built or constructed. [**70:**100]

Switch, Isolating. A switch intended for isolating an electric circuit from the source of power. It has no interrupting rating, and it is intended to be operated only after the circuit has been opened by some other means. [**70:**100]

Switchboard. A large single panel, frame, or assembly of panels on which are mounted on the face, back, or both, switches, overcurrent and other protective devices, buses, and usually instruments. These assemblies are generally accessible from the rear as well as from the front and are not intended to be installed in cabinets. [**70:**100]

Switchgear, Arc-Resistant. Equipment designed to withstand the effects of an internal arcing fault and that directs the internally released energy away from the employee.

Switchgear, Metal-Clad. A switchgear assembly completely enclosed on all sides and top with sheet metal, having drawout switching and interrupting devices, and all live parts enclosed within grounded metal compartments.

Switchgear, Metal-Enclosed. A switchgear assembly completely enclosed on all sides and top with sheet metal (except for ventilating openings and inspection windows), containing primary power circuit switching, interrupting devices, or both, with buses and connections. This assembly may include control and auxiliary devices. Access to the interior of the enclosure is provided by doors, removable covers, or both. Metal-enclosed switchgear is available in non-arc-resistant or arc-resistant constructions.

Switching Device. A device designed to close, open, or both, one or more electric circuits.

Touch Potential. A ground potential gradient difference that can cause current flow from hand to hand, hand to foot, or another path, other than foot to foot, through the body.

Ungrounded. Not connected to ground or to a conductive body that extends the ground connection. [**70:**100]

Unqualified Person. A person who is not a qualified person.

Utilization Equipment. Equipment that utilizes electric energy for electronic, electromechanical, chemical, heating, lighting, or similar purposes. [**70:**100]

Voltage (of a Circuit). The greatest root-mean-square (rms) (effective) difference of potential between any two conductors of the circuit concerned. [**70:**100]

Informational Note: Some systems, such as three-phase 4-wire, single-phase 3-wire, and 3-wire direct-current, may have various circuits of various voltages.

Voltage, Nominal. A nominal value assigned to a circuit or system for the purpose of conveniently designating its voltage class (e.g., 120/240 volts, 480Y/277 volts, 600 volts). [**70:**100]

Informational Note No. 1: The actual voltage at which a circuit operates can vary from the nominal within a range that permits satisfactory operation of equipment.

Informational Note No. 2: See ANSI C84.1, *Electric Power Systems and Equipment — Voltage Ratings (60 Hz).*

Working On (energized electrical conductors or circuit parts). Intentionally coming in contact with energized electrical conductors or circuit parts with the hands, feet, or other body parts, with tools, probes, or with test equipment, regardless of the personal protective equipment (PPE) a person is wearing. There are two categories of "working on": *Diagnostic (testing)* is taking readings or measurements of electrical equipment with approved test equipment that does not require making any physical change to the equipment; *repair* is any physical alteration of electrical equipment (such as making or tightening connections, removing or replacing components, etc.).

ARTICLE 105
Application of Safety-Related Work Practices

105.1 Scope. Chapter 1 covers electrical safety-related work practices and procedures for employees who are exposed to an electrical hazard in workplaces covered in the scope of this standard.

105.2 Purpose. These practices and procedures are intended to provide for employee safety relative to electrical hazards in the workplace.

Informational Note: For general categories of electrical hazards, see Informative Annex K.

105.3 Responsibility. The employer shall provide the safety-related work practices and shall train the employee, who shall then implement them.

105.4 Organization. Chapter 1 of this standard is divided into five articles. Article 100 provides definitions for terms used in one or more of the chapters of this document. Article 105 provides for application of safety-related work practices. Article 110 provides general requirements for electrical safety-related work practices. Article 120 provides requirements for establishing an electrically safe work condition. Article 130 provides requirements for work involving electrical hazards.

ARTICLE 110
General Requirements for Electrical Safety-Related Work Practices

• **110.1 Electrical Safety Program.**

(A) General. The employer shall implement and document an overall electrical safety program that directs activity appropriate to the risk associated with electrical hazards. The electrical safety program shall be implemented as part of the employer's overall occupational health and safety management system, when one exists.

Informational Note No. 1: Safety-related work practices such as verification of proper maintenance and installation, alerting techniques, auditing requirements, and training requirements provided in this standard are administrative controls and part of an overall electrical safety program.

Informational Note No. 2: ANSI/AIHA Z10, *American National Standard for Occupational Health and Safety Management Systems*, provides a framework for establishing a comprehensive electrical safety program as a component of an employer's occupational safety and health program.

Informational Note No. 3: IEEE 3007.1, *Recommended Practice for the Operation and Management of Industrial and Commercial Power Systems*, provides additional guidance for the implementation of the electrical safety program.

Informational Note No. 4: IEEE 3007.3, *Recommended Practice for Electrical Safety in Industrial and Commercial Power Systems*, provides additional guidance for electrical safety in the workplace.

(B) Maintenance. The electrical safety program shall include elements that consider condition of maintenance of electrical equipment and systems.

(C) Awareness and Self-Discipline. The electrical safety program shall be designed to provide an awareness of the potential electrical hazards to employees who work in an environment with the presence of electrical hazards. The program shall be developed to provide the required self-discipline for all employees who must perform work that may involve electrical hazards. The program shall instill safety principles and controls.

(D) Electrical Safety Program Principles. The electrical safety program shall identify the principles upon which it is based.

Informational Note: For examples of typical electrical safety program principles, see Informative Annex E.

(E) Electrical Safety Program Controls. An electrical safety program shall identify the controls by which it is measured and monitored.

Informational Note: For examples of typical electrical safety program controls, see Informative Annex E.

(F) Electrical Safety Program Procedures. An electrical safety program shall identify the procedures to be utilized before work is started by employees exposed to an electrical hazard.

Informational Note: For an example of a typical electrical safety program procedure, see Informative Annex E.

(G) Risk Assessment Procedure. An electrical safety program shall include a risk assessment procedure that addresses employee exposure to electrical hazards. The procedure shall identify the process to be used by the employee before work is started to carry out the following:

(1) Identify hazards
(2) Assess risks
(3) Implement risk control according to a hierarchy of methods

Informational Note No. 1: The hierarchy of risk control methods specified in ANSI/AIHA Z10, *American National Standard for Occupational Health and Safety Management Systems*, is as follows:

(1) Elimination
(2) Substitution
(3) Engineering controls
(4) Awareness
(5) Administrative controls
(6) PPE

Informational Note No. 2: The risk assessment procedure may include identifying when a second person could be required and the training and equipment that person should have.

Informational Note: For an example of a risk assessment procedure, see Informative Annex F.

(H) Job Briefing. Before starting each job, the employee in charge shall conduct a job briefing with the employees involved. The briefing shall cover such subjects as hazards associated with the job, work procedures involved, special precautions, energy source controls, PPE requirements, and the information on the energized electrical work permit, if required. Additional job briefings shall be held if changes that might affect the safety of employees occur during the course of the work.

Informational Note: For an example of a job briefing form and planning checklist, see Figure I.1.

(I) Electrical Safety Auditing.

(1) Electrical Safety Program. The electrical safety program shall be audited to verify that the principles and procedures of the electrical safety program are in compliance with this standard. Audits shall be performed at intervals not to exceed 3 years.

(2) Field Work. Field work shall be audited to verify that the requirements contained in the procedures of the electrical safety program are being followed. When the auditing determines that the principles and procedures of the electrical safety program are not being followed, the appropriate revisions to the training program or revisions to the procedures shall be made. Audits shall be performed at intervals not to exceed 1 year.

(3) Documentation. The audits shall be documented.

110.2 Training Requirements.

(A) Safety Training. The training requirements contained in this section shall apply to employees exposed to an electrical hazard when the risk associated with that hazard is not reduced to a safe level by the applicable electrical installation requirements. Such employees shall be trained to understand the specific hazards associated with electrical energy. They shall be trained in safety-related work practices and procedural requirements, as necessary, to provide protection from the electrical hazards associated with their respective job or task assignments. Employees shall be trained to identify and understand the relationship between electrical hazards and possible injury.

Informational Note: For further information concerning installation requirements, see *NFPA 70, National Electrical Code.*

(B) Type of Training. The training required by this section shall be classroom, on-the-job, or a combination of the two. The type and extent of the training provided shall be determined by the risk to the employee.

(C) Emergency Response Training.

(1) Contact Release. Employees exposed to shock hazards shall be trained in methods of safe release of victims from contact with exposed energized electrical conductors or circuit parts. Refresher training shall occur annually.

(2) First Aid, Emergency Response, and Resuscitation.

(a) Employees responsible for responding to medical emergencies shall be trained in first aid and emergency procedures.

(b) Employees responsible for responding to medical emergencies shall be trained in cardiopulmonary resuscitation (CPR). Refresher training shall occur annually.

(c) Employees responsible for responding to medical emergencies shall be trained in the use of an automated external defibrillator (AED) if an employer's emergency response plan includes the use of this device. Refresher training shall occur annually.

(3) Training Verification. Employers shall verify at least annually that employee training required by this section is current.

(4) Documentation. The employer shall document that the training required by this section has occurred.

(D) Employee Training.

(1) Qualified Person. A qualified person shall be trained and knowledgeable in the construction and operation of equipment or a specific work method and be trained to identify and avoid the electrical hazards that might be present with respect to that equipment or work method.

(a) Such persons shall also be familiar with the proper use of the special precautionary techniques, applicable electrical policies and procedures, PPE, insulating and shielding materials, and insulated tools and test equipment. A person can be considered qualified with respect to certain equipment and methods but still be unqualified for others.

(b) Such persons permitted to work within the limited approach boundary shall, at a minimum, be additionally trained in all of the following:

(1) Skills and techniques necessary to distinguish exposed energized electrical conductors and circuit parts from other parts of electrical equipment
(2) Skills and techniques necessary to determine the nominal voltage of exposed energized electrical conductors and circuit parts
(3) Approach distances specified in Table 130.4(D)(a) and Table 130.4(D)(b) and the corresponding voltages to which the qualified person will be exposed
(4) Decision-making process necessary to be able to do the following:
    a. Perform the job safety planning
    b. Identify electrical hazards
    c. Assess the associated risk
    d. Select the appropriate risk control methods from the hierarchy of controls identified in 110.1(G), including personal protective equipment

(c) An employee who is undergoing on-the-job training for the purpose of obtaining the skills and knowledge necessary to be considered a qualified person, and who in the course of such training demonstrates an ability to perform specific duties safely at his or her level of training, and who is under the direct supervision of a qualified person shall be considered to be a qualified person for the performance of those specific duties.

(d) Tasks that are performed less often than once per year shall require retraining before the performance of the work practices involved.

(e) Employees shall be trained to select an appropriate test instrument and shall demonstrate how to use a device to verify the absence of voltage, including interpreting indications provided by the device. The training shall include information that enables the employee to understand all limitations of each test instrument that might be used.

(f) The employer shall determine through regular supervision or through inspections conducted on at least an annual basis that each employee is complying with the safety-related work practices required by this standard.

(2) Unqualified Persons. Unqualified persons shall be trained in, and be familiar with, any electrical safety-related practices necessary for their safety.

(3) Retraining. Retraining in safety-related work practices and applicable changes in this standard shall be performed at intervals not to exceed three years. An employee shall receive additional training (or retraining) if any of the following conditions exists:

(1) The supervision or annual inspections indicate that the employee is not complying with the safety-related work practices.
(2) New technology, new types of equipment, or changes in procedures necessitate the use of safety-related work practices that are different from those that the employee would normally use.
(3) The employee must employ safety-related work practices that are not normally used during his or her regular job duties.

(E) Training Documentation. The employer shall document that each employee has received the training required by 110.2(D). This documentation shall be made when the employee demonstrates proficiency in the work practices involved and shall be maintained for the duration of the employee's employment. The documentation shall contain the content of the training, each employee's name, and dates of training.

Informational Note No. 1: Content of the training could include one or more of the following: course syllabus, course curriculum, outline, table of contents or training objectives.

Informational Note No. 2: Employment records that indicate that an employee has received the required training are an acceptable means of meeting this requirement.

110.3 Host and Contract Employers' Responsibilities.

(A) Host Employer Responsibilities.

(1) The host employer shall inform contract employers of the following:

(1) Known hazards that are covered by this standard, that are related to the contract employer's work, and that might not be recognized by the contract employer or its employees
(2) Information about the employer's installation that the contract employer needs to make the assessments required by Chapter 1

(2) The host employer shall report observed contract employer–related violations of this standard to the contract employer.

(B) Contract Employer Responsibilities.

(1) The contract employer shall ensure that each of his or her employees is instructed in the hazards communicated to the contract employer by the host employer. This instruction shall be in addition to the basic training required by this standard.

(2) The contract employer shall ensure that each of his or her employees follows the work practices required by this standard and safety-related work rules required by the host employer.

(3) The contract employer shall advise the host employer of the following:

(1) Any unique hazards presented by the contract employer's work
(2) Hazards identified during the course of work by the contract employer that were not communicated by the host employer
(3) The measures the contractor took to correct any violations reported by the host employer under 110.3(A)(2) and to prevent such violation from recurring in the future

(C) Documentation. Where the host employer has knowledge of hazards covered by this standard that are related to the contract employer's work, there shall be a documented meeting between the host employer and the contract employer.

110.4 Use of Electrical Equipment.

(A) Test Instruments and Equipment.

(1) Testing. Only qualified persons shall perform tasks such as testing, troubleshooting, and voltage measuring within the limited approach boundary of energized electrical conductors or circuit parts operating at 50 volts or more or where an electrical hazard exists.

(2) Rating. Test instruments, equipment, and their accessories shall be rated for circuits and equipment where they are utilized.

Informational Note: See ANSI/ISA-61010-1 (82.02.01)/UL 61010-1, *Safety Requirements for Electrical Equipment for Measurement, Control, and Laboratory Use – Part 1: General Requirements*, for rating and design requirements for voltage measurement and test instruments intended for use on electrical systems 1000 volts and below.

(3) Design. Test instruments, equipment, and their accessories shall be designed for the environment to which they will be exposed and for the manner in which they will be utilized.

(4) Visual Inspection and Repair. Test instruments and equipment and all associated test leads, cables, power cords, probes, and connectors shall be visually inspected for external defects and damage before each use. If there is a defect or evidence of damage that might expose an employee to injury, the defective or damaged item shall be removed from service. No employee shall use it until a person(s) qualified to perform the repairs and tests that are necessary to render the equipment safe has done so.

(5) Operation Verification. When test instruments are used for testing the absence of voltage on conductors or circuit parts operating at 50 volts or more, the operation of the test instrument shall be verified on a known voltage source before and after an absence of voltage test is performed.

(B) Portable Electric Equipment. This section applies to the use of cord- and plug-connected equipment, including cord sets (extension cords).

(1) Handling and Storage. Portable equipment shall be handled and stored in a manner that will not cause damage. Flexible electric cords connected to equipment shall not be used for raising or lowering the equipment. Flexible cords shall not be fastened with staples or hung in such a fashion as could damage the outer jacket or insulation.

(2) Grounding-Type Equipment.

(a) A flexible cord used with grounding-type utilization equipment shall contain an equipment grounding conductor.

(b) Attachment plugs and receptacles shall not be connected or altered in a manner that would interrupt continuity of the equipment grounding conductor.

Additionally, these devices shall not be altered in order to allow use in a manner that was not intended by the manufacturer.

(c) Adapters that interrupt the continuity of the equipment grounding conductor shall not be used.

(3) Visual Inspection and Repair of Portable Cord- and Plug-Connected Equipment and Flexible Cord Sets.

(a) Frequency of Inspection. Before each use, portable cord- and plug-connected equipment shall be visually inspected for external defects (such as loose parts or deformed and missing pins) and for evidence of possible internal damage (such as a pinched or crushed outer jacket).

Exception: Cord- and plug-connected equipment and flexible cord sets (extension cords) that remain connected once they are put in place and are not exposed to damage shall not be required to be visually inspected until they are relocated.

(b) Defective Equipment. If there is a defect or evidence of damage that might expose an employee to injury, the defective or damaged item shall be removed from service. No employee shall use it until a person(s) qualified to perform the repairs and tests necessary to render the equipment safe has done so.

(c) Proper Mating. When an attachment plug is to be connected to a receptacle, the relationship of the plug and receptacle contacts shall first be checked to ensure that they are of mating configurations.

(4) Conductive Work Locations. Portable electric equipment used in highly conductive work locations (such as those inundated with water or other conductive liquids) shall be approved for those locations. In job locations where employees are likely to contact or be drenched with water or conductive liquids, ground-fault circuit-interrupter protection for personnel shall also be used.

Informational Note: The risk assessment procedure can also include identifying when the use of portable tools and equipment powered by sources other than 120 volts ac, such as batteries, air, and hydraulics, should be used to minimize the potential for injury from electrical hazards for tasks performed in conductive or wet locations.

(5) Connecting Attachment Plugs.

(a) Employees' hands shall not be wet when plugging and unplugging flexible cords and cord- and plug-connected equipment if energized equipment is involved.

(b) Energized plug and receptacle connections shall be handled only with insulating protective equipment if the condition of the connection could provide a conductive path to the employee's hand (e.g, if a cord connector is wet from being immersed in water).

(c) Locking-type connectors shall be secured after connection.

(6) Manufacturer's Instructions. Portable equipment shall be used in accordance with the manufacturer's instructions and safety warnings.

(C) Ground-Fault Circuit-Interrupter (GFCI) Protection.

(1) General. Employees shall be provided with ground-fault circuit-interrupter (GFCI) protection where required by applicable state, federal, or local codes and standards. Listed cord sets or devices incorporating listed GFCI protection for personnel identified for portable use shall be permitted.

(2) Maintenance and Construction. GFCI protection shall be provided where an employee is operating or using cord- and plug-connected tools related to maintenance and construction activity supplied by 125-volt, 15-, 20-, or 30-ampere circuits. Where employees operate or use equipment supplied by greater than 125-volt, 15-, 20-, or 30-ampere circuits, GFCI protection or an assured equipment grounding conductor program shall be implemented.

(3) Outdoors. GFCI protection shall be provided when an employee is outdoors and operating or using cord- and plug-connected equipment supplied by 125-volt, 15-, 20-, or 30-ampere circuits. Where employees working outdoors operate or use equipment supplied by greater than 125-volt, 15-, 20-, or 30-ampere circuits, GFCI protection or an assured equipment grounding conductor program shall be implemented.

(D) Ground-Fault Circuit-Interrupter Protection Devices. GFCI protection devices shall be tested in accordance with the manufacturer's instructions.

(E) Overcurrent Protection Modification. Overcurrent protection of circuits and conductors shall not be modified, even on a temporary basis, beyond what is permitted by applicable portions of electrical codes and standards dealing with overcurrent protection.

Informational Note: For further information concerning electrical codes and standards dealing with overcurrent protection, refer to Article 240 of *NFPA 70, National Electrical Code*.

•

ARTICLE 120
Establishing an Electrically Safe Work Condition

120.1 Verification of an Electrically Safe Work Condition. An electrically safe work condition shall be achieved when performed in accordance with the procedures of 120.2 and verified by the following process:

(1) Determine all possible sources of electrical supply to the specific equipment. Check applicable up-to-date drawings, diagrams, and identification tags.
(2) After properly interrupting the load current, open the disconnecting device(s) for each source.
(3) Wherever possible, visually verify that all blades of the disconnecting devices are fully open or that drawout-type circuit breakers are withdrawn to the fully disconnected position.
(4) Apply lockout/tagout devices in accordance with a documented and established policy.
(5) Use an adequately rated test instrument to test each phase conductor or circuit part to verify it is de-energized. Test each phase conductor or circuit part both phase-to-phase and phase-to-ground. Before and after each test, determine that the test instrument is operating satisfactorily through verification on a known voltage source.

Informational Note: See ANSI/ISA 61010-1, *Safety Requirements for Electrical Equipment for Measurement, Control, and Laboratory Use, Part 1: General Requirements*, for rating and design requirements for voltage measurement and test instruments intended for use on electrical systems 1000 volts and below.

(6) Where the possibility of induced voltages or stored electrical energy exists, ground the phase conductors or circuit parts before touching them. Where it could be reasonably anticipated that the conductors or circuit parts being de-energized could contact other exposed energized conductors or circuit parts, apply ground connecting devices rated for the available fault duty.

120.2 De-energized Electrical Equipment That Has Lockout/Tagout Devices Applied. Each employer shall identify, document, and implement lockout/tagout procedures conforming to Article 120 to safeguard employees from exposure to electrical hazards. The lockout/tagout procedure shall be appropriate for the experience and training of the employees and conditions as they exist in the workplace.

(A) General. All electrical circuit conductors and circuit parts shall be considered energized until the source(s) of energy is (are) removed, at which time they shall be considered de-energized. All electrical conductors and circuit parts shall not be considered to be in an electrically safe work condition until all of the applicable requirements of Article 120 have been met.

Informational Note: See 120.1 for the six-step procedure to verify an electrically safe work condition.

Electrical conductors and circuit parts that have been disconnected, but not under lockout/tagout; tested; and grounded (where appropriate) shall not be considered to be in an electrically safe work condition, and safe work practices appropriate for the circuit voltage and energy level shall be used. Lockout/tagout requirements shall apply to fixed, permanently installed equipment; to temporarily installed equipment; and to portable equipment.

(B) Principles of Lockout/Tagout Execution.

(1) Employee Involvement. Each person who could be exposed directly or indirectly to a source of electrical energy shall be involved in the lockout/tagout process.

Informational Note: An example of direct exposure is the qualified electrician who works on the motor starter control, the power circuits, or the motor. An example of indirect exposure is the person who works on the coupling between the motor and compressor.

(2) Training. All persons who could be exposed or affected by the lockout/tagout shall be trained to understand the established procedure to control the energy and their responsibility in the procedure and its execution. New or reassigned employees shall be trained to understand the lockout/tagout procedure as it relates to their new assignments.

(3) Retraining. Retraining shall be performed:

(a) When the established procedure is revised

(b) At intervals not to exceed 3 years

(4) Training Documentation.

(a) The employer shall document that each employee has received the training required by this section.

(b) The documentation shall be made when the employee demonstrates proficiency in the work practices involved.

(c) The documentation shall contain the content of the training, each employee's name, and the dates of the training.

Informational Note: Content of the training could include one or more of the following: course syllabus, course curriculum, outline, table of contents, or training objectives.

(5) Plan. A plan shall be developed on the basis of the existing electrical equipment and system and shall use up-to-date diagrammatic drawing representation(s).

(6) Control of Energy. All sources of electrical energy shall be controlled in such a way as to minimize employee exposure to electrical hazards.

(7) Identification. The lockout/tagout device shall be unique and readily identifiable as a lockout/tagout device.

(8) Voltage. Voltage shall be removed and absence of voltage verified.

(9) Coordination. The established electrical lockout/tagout procedure shall be coordinated with all of the employer's procedures associated with lockout/tagout of other energy sources.

(C) Responsibility.

(1) Procedures. The employer shall establish lockout/tagout procedures for the organization, provide training to employees, provide equipment necessary to execute the details of the procedure, audit execution of the procedures to ensure employee understanding/compliance, and audit the procedure for improvement opportunity and completeness.

(2) Form of Control. Two forms of hazardous electrical energy control shall be permitted: simple lockout/tagout and complex lockout/tagout *[see 120.2(D)]*. For the simple lockout/tagout, the qualified person shall be in charge. For the complex lockout/tagout, the person in charge shall have overall responsibility.

Informational Note: For an example of a lockout/tagout procedure, see Informative Annex G.

(3) Audit Procedures. An audit shall be conducted at least annually by a qualified person and shall cover at least one lockout/tagout in progress and the procedure details. The audit shall be designed to correct deficiencies in the established electrical lockout/tagout procedure or in employee understanding.

(D) Hazardous Electrical Energy Control Procedure.

(1) Simple Lockout/Tagout Procedure. All lockout/tagout procedures that involve only a qualified person(s) deenergizing one set of conductors or circuit part source for the sole purpose of safeguarding employees from exposure to electrical hazards shall be considered to be a simple lockout/tagout. Simple lockout/tagout plans shall not be required to be written for each application. Each worker shall be responsible for his or her own lockout/tagout.

(2) Complex Lockout/Tagout Procedure.

(a) A complex lockout/tagout plan shall be permitted where one or more of the following exist:

(1) Multiple energy sources
(2) Multiple crews
(3) Multiple crafts
(4) Multiple locations
(5) Multiple employers
(6) Multiple disconnecting means
(7) Particular sequences
(8) Job or task that continues for more than one work period

(b) All complex lockout/tagout procedures shall require a written plan of execution that identifies the person in charge.

(c) The complex lockout/tagout procedure shall vest primary responsibility in an authorized employee for a set number of employees working under the protection of a group lockout or tagout device (such as an operation lock). The person in charge shall be held accountable for safe execution of the complex lockout/tagout.

(d) Each authorized employee shall affix a personal lockout or tagout device to the group lockout device, group lockbox, or comparable mechanism when he or she begins work and shall remove those devices when he or she stops working on the machine or equipment being serviced or maintained.

(e) The complex lockout/tagout procedure shall address all the concerns of employees who might be exposed. All complex lockout/tagout plans shall identify the method to account for all persons who might be exposed to electrical hazards in the course of the lockout/tagout.

(3) Coordination.

(a) The established electrical lockout/tagout procedure shall be coordinated with all other employer's procedures for control of exposure to electrical energy sources such that all employer's procedural requirements are adequately addressed on a site basis.

(b) The procedure for control of exposure to electrical hazards shall be coordinated with other procedures for control of other hazardous energy sources such that they are based on similar/identical concepts.

(c) The electrical lockout/tagout procedure shall always include voltage testing requirements where there might be direct exposure to electrical hazards.

(d) Electrical lockout/tagout devices shall be permitted to be similar to lockout/tagout devices for control of other hazardous energy sources, such as pneumatic, hydraulic, thermal, and mechanical, if such devices are used only for control of hazardous energy and for no other purpose.

(E) Equipment.

(1) Lock Application. Energy isolation devices for machinery or equipment installed after January 2, 1990, shall be capable of accepting a lockout device.

(2) Lockout/Tagout Device. Each employer shall supply, and employees shall use, lockout/tagout devices and equipment necessary to execute the requirements of 120.2(E). Locks and tags used for control of exposure to electrical hazards shall be unique, shall be readily identifiable as lockout/tagout devices, and shall be used for no other purpose.

(3) Lockout Device.

(a) A lockout device shall include a lock (either keyed or combination).

(b) The lockout device shall include a method of identifying the individual who installed the lockout device.

(c) A lockout device shall be permitted to be only a lock, if the lock is readily identifiable as a lockout device, in addition to having a means of identifying the person who installed the lock.

(d) Lockout devices shall be attached to prevent operation of the disconnecting means without resorting to undue force or the use of tools.

(e) Where a tag is used in conjunction with a lockout device, the tag shall contain a statement prohibiting unauthorized operation of the disconnecting means or unauthorized removal of the device.

(f) Lockout devices shall be suitable for the environment and for the duration of the lockout.

(g) Whether keyed or combination locks are used, the key or combination shall remain in the possession of the individual installing the lock or the person in charge, when provided by the established procedure.

(4) Tagout Device.

(a) A tagout device shall include a tag together with an attachment means.

(b) The tagout device shall be readily identifiable as a tagout device and suitable for the environment and duration of the tagout.

(c) A tagout device attachment means shall be capable of withstanding at least 224.4 N (50 lb) of force exerted at a right angle to the disconnecting means surface. The tag attachment means shall be nonreusable, attachable by hand, self-locking, nonreleasable, and equal to an all-environmental tolerant nylon cable tie.

(d) Tags shall contain a statement prohibiting unauthorized operation of the disconnecting means or removal of the tag.

(e) A hold card tagging tool on an overhead conductor in conjunction with a hotline tool to install the tagout device safely on a disconnect that is isolated from the work(s) shall be permitted. Where a hold card is used, the tagout procedure shall include the method of accounting for personnel who are working under the protection of the hold card.

(5) Electrical Circuit Interlocks. Up-to-date diagrammatic drawings shall be consulted to ensure that no electrical circuit interlock operation can result in reenergizing the circuit being worked on.

(6) Control Devices. Locks/tags shall be installed only on circuit disconnecting means. Control devices, such as pushbuttons or selector switches, shall not be used as the primary isolating device.

(F) Procedures. The employer shall maintain a copy of the procedures required by this section and shall make the procedures available to all employees.

(1) Planning. The procedure shall require planning, including the requirements of 120.2(F)(1)(a) through 120.2(F)(2)(n).

(a) Locating Sources. Up-to-date single-line drawings shall be considered a primary reference source for such information. When up-to-date drawings are not available, the employer shall be responsible for ensuring that an equally effective means of locating all sources of energy is employed.

(b) Exposed Persons. The plan shall identify persons who might be exposed to an electrical hazard and the PPE required during the execution of the job or task.

(c) Person In Charge. The plan shall identify the person in charge and his or her responsibility in the lockout/tagout.

(d) Simple Lockout/Tagout. Simple lockout/tagout procedure shall be in accordance with 120.2(D)(1).

(e) Complex Lockout/Tagout. Complex lockout/tagout procedure shall be in accordance with 120.2(D)(2).

(2) Elements of Control. The procedure shall identify elements of control.

(a) De-energizing Equipment (Shutdown). The procedure shall establish the person who performs the switching and where and how to de-energize the load.

(b) Stored Energy. The procedure shall include requirements for releasing stored electric or mechanical energy that might endanger personnel. All capacitors shall be discharged, and high capacitance elements shall also be short-circuited and grounded before the associated equipment is touched or worked on. Springs shall be released or physical restraint shall be applied when necessary to immobilize mechanical equipment and pneumatic and hydraulic pressure reservoirs. Other sources of stored energy shall be blocked or otherwise relieved.

(c) Disconnecting Means. The procedure shall identify how to verify that the circuit is de-energized (open).

(d) Responsibility. The procedure shall identify the person who is responsible for verifying that the lockout/tagout procedure is implemented and who is responsible for ensuring that the task is completed prior to removing locks/tags. A mechanism to accomplish lockout/tagout for multiple (complex) jobs/tasks where required, including the person responsible for coordination, shall be included.

(e) Verification. The procedure shall verify that equipment cannot be restarted. The equipment operating controls, such as pushbuttons, selector switches, and electrical interlocks, shall be operated or otherwise it shall be verified that the equipment cannot be restarted.

(f) Testing. The procedure shall establish the following:

(1) Voltage detector to be used, the required PPE, and the person who will use it to verify proper operation of the voltage detector before and after use
(2) Requirement to define the boundary of the electrically safe work condition
(3) Requirement to test before touching every exposed conductor or circuit part(s) within the defined boundary of the work area
(4) Requirement to retest for absence of voltage when circuit conditions change or when the job location has been left unattended
(5) Planning considerations that include methods of verification where there is no accessible exposed point to take voltage measurements

(g) Grounding. Grounding requirements for the circuit shall be established, including whether the temporary protective grounding equipment shall be installed for the duration of the task or is temporarily established by the procedure. Grounding needs or requirements shall be permitted to be covered in other work rules and might not be part of the lockout/tagout procedure.

(h) Shift Change. A method shall be identified in the procedure to transfer responsibility for lockout/tagout to another person or to the person in charge when the job or task extends beyond one shift.

(i) Coordination. The procedure shall establish how coordination is accomplished with other jobs or tasks in progress, including related jobs or tasks at remote locations, including the person responsible for coordination.

(j) Accountability for Personnel. A method shall be identified in the procedure to account for all persons who could be exposed to hazardous energy during the lockout/tagout.

(k) Lockout/Tagout Application. The procedure shall clearly identify when and where lockout applies, in addition to when and where tagout applies, and shall address the following:

(1) Lockout shall be defined as installing a lockout device on all sources of hazardous energy such that operation of the disconnecting means is prohibited and forcible removal of the lock is required to operate the disconnecting means.
(2) Tagout shall be defined as installing a tagout device on all sources of hazardous energy, such that operation of the disconnecting means is prohibited. The tagout device shall be installed in the same position available for the lockout device.
(3) Where it is not possible to attach a lock to existing disconnecting means, the disconnecting means shall not be used as the only means to put the circuit in an electrically safe work condition.
(4) The use of tagout procedures without a lock shall be permitted only in cases where equipment design precludes the installation of a lock on an energy isolation device(s). When tagout is employed, at least one additional safety measure shall be employed. In such cases, the procedure shall clearly establish responsibilities and accountability for each person who might be exposed to electrical hazards.

Informational Note: Examples of additional safety measures include the removal of an isolating circuit element such as fuses, blocking of the controlling switch, or opening an extra disconnecting device to reduce the likelihood of inadvertent energization.

(l) Removal of Lockout/Tagout Devices. The procedure shall identify the details for removing locks or tags when the installing individual is unavailable. When locks or tags are removed by someone other than the installer, the employer shall attempt to locate that person prior to removing the lock or tag. When the lock or tag is removed because the installer is unavailable, the installer shall be informed prior to returning to work.

(m) Release for Return to Service. The procedure shall identify steps to be taken when the job or task requiring lockout/tagout is completed. Before electric circuits or equipment are reenergized, appropriate tests and visual inspections shall be conducted to verify that all tools, mechanical restraints and electrical jumpers, short circuits, and temporary protective grounding equipment have been removed, so that the circuits and equipment are in a condition to be safely energized. Where appropriate, the employees responsible for operating the machines or process shall be notified when circuits and equipment are ready to be energized, and such employees shall provide assistance as necessary to safely energize the circuits and equipment. The procedure shall contain a statement requiring the area to be inspected to ensure that nonessential items have been removed. One such step shall ensure that all personnel are clear of exposure to dangerous conditions resulting from reenergizing the service and that blocked mechanical equipment or grounded equipment is cleared and prepared for return to service.

(n) Temporary Release for Testing/Positioning. The procedure shall clearly identify the steps and qualified persons' responsibilities when the job or task requiring lockout/tagout is to be interrupted temporarily for testing or positioning of equipment; then the steps shall be identical to the steps for return to service.

Informational Note: See 110.4(A) for requirements when using test instruments and equipment.

120.3 Temporary Protective Grounding Equipment.

(A) Placement. Temporary protective grounding equipment shall be placed at such locations and arranged in such a manner as to prevent each employee from being exposed to a shock hazard (hazardous differences in electrical potential). The location, sizing, and application of temporary protective grounding equipment shall be identified as part of the employer's job planning.

(B) Capacity. Temporary protective grounding equipment shall be capable of conducting the maximum fault current that could flow at the point of grounding for the time necessary to clear the fault.

(C) Equipment Approval. Temporary protective grounding equipment shall meet the requirements of ASTM F855, *Standard Specification for Temporary Protective Grounds to be Used on De-energized Electric Power Lines and Equipment*.

(D) Impedance. Temporary protective grounding equipment and connections shall have an impedance low enough to cause immediate operation of protective devices in case of accidental energizing of the electric conductors or circuit parts.

ARTICLE 130
Work Involving Electrical Hazards

130.1 General. Article 130 covers the following:

(1) When an electrically safe work condition must be established

(2) The electrical safety-related work practices when an electrically safe work condition cannot be established

All requirements of this article shall apply whether an incident energy analysis is completed or if Table 130.7(C)(15)(A)(a), Table 130.7(C)(15)(A)(b), Table 130.7(C)(15)(B), and Table 130.7(C)(16) are used in lieu of an incident energy analysis in accordance with 130.5.

130.2 Electrically Safe Working Conditions. Energized electrical conductors and circuit parts shall be put into an electrically safe work condition before an employee performs work if any of the following conditions exist:

(1) The employee is within the limited approach boundary.
(2) The employee interacts with equipment where conductors or circuit parts are not exposed but an increased likelihood of injury from an exposure to an arc flash hazard exists.

Exception: Where a disconnecting means or isolating element that has been properly installed and maintained is operated, opened, closed, removed, or inserted to achieve an electrically safe work condition for connected equipment or to return connected equipment to service that has been placed in an electrically safe work condition, the equipment supplying the disconnecting means or isolating element shall not be required to be placed in an electrically safe work condition provided a risk assessment is performed and does not identify unacceptable risks for the task.

(A) Energized Work.

(1) Additional Hazards or Increased Risk. Energized work shall be permitted where the employer can demonstrate that de-energizing introduces additional hazards or increased risk.

(2) Infeasibility. Energized work shall be permitted where the employer can demonstrate that the task to be performed is infeasible in a de-energized state due to equipment design or operational limitations.

(3) Less Than 50 Volts. Energized electrical conductors and circuit parts that operate at less than 50 volts shall not be required to be de-energized where the capacity of the source and any overcurrent protection between the energy source and the worker are considered and it is determined that there will be no increased exposure to electrical burns or to explosion due to electric arcs.

Informational Note No. 1: Examples of additional hazards or increased risk include, but are not limited to, interruption of life-support equipment, deactivation of emergency alarm systems, and shutdown of hazardous location ventilation equipment.

Informational Note No. 2: Examples of work that might be performed within the limited approach boundary of exposed energized electrical conductors or circuit parts because of infeasibility due to equipment design or operational limitations include performing diagnostics and testing (for example, start-up or troubleshooting) of electric circuits that can only be performed with the circuit energized and work on circuits that form an integral part of a continuous process that would otherwise need to be completely shut down in order to permit work on one circuit or piece of equipment.

(4) Normal Operation. Normal operation of electric equipment shall be permitted where all of the following conditions are satisfied:

(1) The equipment is properly installed.
(2) The equipment is properly maintained.
(3) The equipment doors are closed and secured.
(4) All equipment covers are in place and secured.
(5) There is no evidence of impending failure.

Informational Note: The phrase *properly installed* means that the equipment is installed in accordance with applicable industry codes and standards and the manufacturer's recommendations. The phrase *properly maintained* means that the equipment has been maintained in accordance with the manufacturer's recommendations and applicable industry codes and standards. The phrase *evidence of impending failure* means that there is evidence such as arcing, overheating, loose or bound equipment parts, visible damage, or deterioration.

(B) Energized Electrical Work Permit.

(1) When Required. When energized work is permitted in accordance with 130.2(A), an energized electrical work permit shall be required under the following conditions:

(1) When work is performed within the restricted approach boundary
(2) When the employee interacts with the equipment when conductors or circuit parts are not exposed but an increased likelihood of injury from an exposure to an arc flash hazard exists

(2) Elements of Work Permit. The energized electrical work permit shall include, but not be limited to, the following items:

(1) Description of the circuit and equipment to be worked on and their location
(2) Justification for why the work must be performed in an energized condition *[see 130.2(A)]*
(3) Description of the safe work practices to be employed *(see 130.3)*
(4) Results of the shock risk assessment *[see 130.4(A)]*
   a. Voltage to which personnel will be exposed
   b. Limited approach boundary *[see 130.4(B), Table 130.4(D)(a), and Table 130.4(D)(b)]*
   c. Restricted approach boundary *[see 130.4(B) and Table 130.4(D)(a) and Table 130.4(D)(b)]*

d. Necessary personal and other protective equipment to safely perform the assigned task *[see 130.4(C), 130.7(C)(1) through (C)(16), Table 130.7(C)(15)(A)(a), Table 130.7(C)(16), and 130.7(D)]*

(5) Results of the arc flash risk assessment *[see 130.5]*

a. Available incident energy at the working distance or arc flash PPE category *[see 130.5]*

b. Necessary PPE to protect against the hazard *[see 130.5(C), 130.7(C)(1) through (C)(16), Table 130.7(C)(15)(A)(a), Table 130.7(C)(16), and 130.7(D)]*

c. Arc flash boundary *[see 130.5(B)]*

(6) Means employed to restrict the access of unqualified persons from the work area *[see 130.3]*

(7) Evidence of completion of a job briefing, including a discussion of any job-specific hazards *(see 130.3)*

(8) Energized work approval (authorizing or responsible management, safety officer, or owner, etc.) signature(s)

Informational Note: For an example of an acceptable energized work permit, see Figure J.1.

(3) Exemptions to Work Permit. An energized electrical work permit shall not be required if a qualified person is provided with and uses appropriate safe work practices and PPE in accordance with Chapter 1 under any of the following conditions:

(1) Testing, troubleshooting, and voltage measuring

(2) Thermography and visual inspections if the restricted approach boundary is not crossed

(3) Access to and egress from an area with energized electrical equipment if no electrical work is performed and the restricted approach boundary is not crossed

(4) General housekeeping and miscellaneous non-electrical tasks if the restricted approach boundary is not crossed

130.3 Working While Exposed to Electrical Hazards. Safety-related work practices shall be used to safeguard employees from injury while they are exposed to electrical hazards from electrical conductors or circuit parts that are or can become energized. The specific safety-related work practices shall be consistent with the electrical hazards and the associated risk. Appropriate safety-related work practices shall be determined before any person is exposed to the electrical hazards involved by using both shock risk assessment and arc flash risk assessment. Only qualified persons shall be permitted to work on electrical conductors or circuit parts that have not been put into an electrically safe work condition.

130.4 Approach Boundaries to Energized Electrical Conductors or Circuit Parts for Shock Protection.

(A) Shock Risk Assessment. A shock risk assessment shall determine the voltage to which personnel will be exposed, the boundary requirements, and the PPE necessary in order to minimize the possibility of electric shock to personnel.

(B) Shock Protection Boundaries. The shock protection boundaries identified as limited approach boundary and restricted approach boundary shall be applicable where approaching personnel are exposed to energized electrical conductors or circuit parts. Table 130.4(D)(a) shall be used for the distances associated with various ac system voltages. Table 130.4(D)(b) shall be used for the distances associated with various dc system voltages.

Informational Note: In certain instances, the arc flash boundary might be a greater distance from the energized electrical conductors or circuit parts than the limited approach boundary. The shock protection boundaries and the arc flash boundary are independent of each other.

(C) Limited Approach Boundary.

(1) Approach by Unqualified Persons. Unless permitted by 130.4(C)(3), no unqualified person shall be permitted to approach nearer than the limited approach boundary of energized conductors and circuit parts.

(2) Working at or Close to the Limited Approach Boundary. Where one or more unqualified persons are working at or close to the limited approach boundary, the designated person in charge of the work space where the electrical hazard exists shall advise the unqualified person(s) of the electrical hazard and warn him or her to stay outside of the limited approach boundary.

(3) Entering the Limited Approach Boundary. Where there is a need for an unqualified person(s) to cross the limited approach boundary, a qualified person shall advise him or her of the possible hazards and continuously escort the unqualified person(s) while inside the limited approach boundary. Under no circumstance shall the escorted unqualified person(s) be permitted to cross the restricted approach boundary.

(D) Restricted Approach Boundary. No qualified person shall approach or take any conductive object closer to exposed energized electrical conductors or circuit parts operating at 50 volts or more than the restricted approach boundary set forth in Table 130.4(D)(a) and Table 130.4(D)(b), unless one of the following conditions applies:

(1) The qualified person is insulated or guarded from the energized electrical conductors or circuit parts operating at 50 volts or more. Insulating gloves or insulating gloves and sleeves are considered insulation only with regard to the energized parts upon which work is being performed. If there is a need for an uninsulated part of the qualified person's body to contact exposed energized electrical conductors or circuit parts, a combination of 130.4(D)(1), 130.4(D)(2), and 130.4(D)(3) shall be used to protect the uninsulated body parts.

(2) The energized electrical conductors or circuit part operating at 50 volts or more are insulated from the qualified person and from any other conductive object at a different potential.

(3) The qualified person is insulated from any other conductive object.

130.5 Arc Flash Risk Assessment. An arc flash risk assessment shall be performed and shall:

(1) Determine if an arc flash hazard exists. If an arc flash hazard exists, the risk assessment shall determine:

a. Appropriate safety-related work practices

b. The arc flash boundary

c. The PPE to be used within the arc flash boundary

(2) Be updated when a major modification or renovation takes place. It shall be reviewed periodically, at intervals not to exceed 5 years, to account for changes in the electrical distribution system that could affect the results of the arc flash risk assessment.

(3) Take into consideration the design of the overcurrent protective device and its opening time, including its condition of maintenance.

Table 130.4(D)(a) Approach Boundaries to Energized Electrical Conductors or Circuit Parts for Shock Protection for Alternating-Current Systems (All dimensions are distance from energized electrical conductor or circuit part to employee.)

(1)	(2)	(3)	(4)
	Limited Approach Boundary[b]		Restricted Approach Boundary[b]; Includes Inadvertent Movement Adder
Nominal System Voltage Range, Phase to Phase[a]	Exposed Movable Conductor[c]	Exposed Fixed Circuit Part	
<50 V	Not specified	Not specified	Not specified
50 V–150 V[d]	3.0 m (10 ft 0 in.)	1.0 m (3 ft 6 in.)	Avoid contact
151 V–750 V	3.0 m (10 ft 0 in.)	1.0 m (3 ft 6 in.)	0.3 m (1 ft 0 in.)
751 V–15 kV	3.0 m (10 ft 0 in.)	1.5 m (5 ft 0 in.)	0.7 m (2 ft 2 in.)
15.1 kV–36 kV	3.0 m (10 ft 0 in.)	1.8 m (6 ft 0 in.)	0.8 m (2 ft 7 in.)
36.1 kV–46 kV	3.0 m (10 ft 0 in.)	2.5 m (8 ft 0 in.)	0.8 m (2 ft 9 in.)
46.1 kV–72.5 kV	3.0 m (10 ft 0 in.)	2.5 m (8 ft 0 in.)	1.0 m (3 ft 3 in.)
72.6 kV–121 kV	3.3 m (10 ft 8 in.)	2.5 m (8 ft 0 in.)	1.0 m (3 ft 4 in.)
138 kV–145 kV	3.4 m (11 ft 0 in.)	3.0 m (10 ft 0 in.)	1.2 m (3 ft 10 in.)
161 kV–169 kV	3.6 m (11 ft 8 in.)	3.6 m (11 ft 8 in.)	1.3 m (4 ft 3 in.)
230 kV–242 kV	4.0 m (13 ft 0 in.)	4.0 m (13 ft 0 in.)	1.7 m (5 ft 8 in.)
345 kV–362 kV	4.7 m (15 ft 4 in.)	4.7 m (15 ft 4 in.)	2.8 m (9 ft 2 in.)
500 kV–550 kV	5.8 m (19 ft 0 in.)	5.8 m (19 ft 0 in.)	3.6 m (11 ft 10 in.)
765 kV–800 kV	7.2 m (23 ft 9 in.)	7.2 m (23 ft 9 in.)	4.9 m (15 ft 11 in.)

Note (1): For arc flash boundary, see 130.5(A).
Note (2): All dimensions are distance from exposed energized electrical conductors or circuit part to employee.
[a] For single-phase systems above 250V, select the range that is equal to the system's maximum phase-to-ground voltage multiplied by 1.732.
[b] See definition in Article 100 and text in 130.4(D)(2) and Informative Annex C for elaboration.
[c] *Exposed movable conductors* describes a condition in which the distance between the conductor and a person is not under the control of the person. The term is normally applied to overhead line conductors supported by poles.
[d] This includes circuits where the exposure does not exceed 120V.

Table 130.4(D)(b) Approach Boundaries to Energized Electrical Conductors or Circuit Parts for Shock Protection, Direct-Current Voltage Systems

(1)	(2)	(3)	(4)
	Limited Approach Boundary		Restricted Approach Boundary; Includes Inadvertent Movement Adder
Nominal Potential Difference	Exposed Movable Conductor*	Exposed Fixed Circuit Part	
<100 V	Not specified	Not specified	Not specified
100 V–300 V	3.0 m (10 ft 0 in.)	1.0 m (3 ft 6 in.)	Avoid contact
301 V–1 kV	3.0 m (10 ft 0 in.)	1.0 m (3 ft 6 in.)	0.3 m (1 ft 0 in.)
1.1 kV–5 kV	3.0 m (10 ft 0 in.)	1.5 m (5 ft 0 in.)	0.5 m (1 ft 5 in.)
5 kV–15 kV	3.0 m (10 ft 0 in.)	1.5 m (5 ft 0 in.)	0.7 m (2 ft 2 in.)
15.1 kV–45 kV	3.0 m (10 ft 0 in.)	2.5 m (8 ft 0 in.)	0.8 m (2 ft 9 in.)
45.1 kV– 75 kV	3.0 m (10 ft 0 in.)	2.5 m (8 ft 0 in.)	1.0 m (3 ft 2 in.)
75.1 kV–150 kV	3.3 m (10 ft 8 in.)	3.0 m (10 ft 0 in.)	1.2 m (4 ft 0 in.)
150.1 kV–250 kV	3.6 m (11 ft 8 in.)	3.6 m (11 ft 8 in.)	1.6 m (5 ft 3 in.)
250.1 kV–500 kV	6.0 m (20 ft 0 in.)	6.0 m (20 ft 0 in.)	3.5 m (11 ft 6 in.)
500.1 kV–800 kV	8.0 m (26 ft 0 in.)	8.0 m (26 ft 0 in.)	5.0 m (16 ft 5 in.)

Note: All dimensions are distance from exposed energized electrical conductors or circuit parts to worker.
* *Exposed movable conductor* describes a condition in which the distance between the conductor and a person is not under the control of the person. The term is normally applied to overhead line conductors supported by poles.

Informational Note No. 1: Improper or inadequate maintenance can result in increased opening time of the overcurrent protective device, thus increasing the incident energy. Where equipment is not properly installed or maintained, PPE selection based on incident energy analysis or the PPE category method may not provide adequate protection from arc flash hazards.

Informational Note No. 2: Both larger and smaller available short-circuit currents could result in higher available arc flash energies. If the available short-circuit current increases without a decrease in the opening time of the overcurrent protective device, the arc flash energy will increase. If the available short-circuit current decreases, resulting in a longer opening time for the overcurrent protective device, arc flash energies could also increase.

Informational Note No. 3: The occurrence of an arcing fault inside an enclosure produces a variety of physical phenomena very different from a bolted fault. For example, the arc energy resulting from an arc developed in the air will cause a sudden pressure increase and localized overheating. Equipment and design practices are available to minimize the energy levels and the number of procedures that could expose an employee to high levels of incident energy. Proven designs such as arc-resistant switchgear, remote racking (insertion or removal), remote opening and closing of switching devices, high-resistance grounding of low-voltage and 5000 volts (nominal) systems, current limitation, and specification of covered bus or covered conductors within equipment are available to reduce the risk associated with an arc flash incident. See Informative Annex O for Safety-Related Design Requirements.

Informational Note No. 4: For additional direction for performing maintenance on overcurrent protective devices, see Chapter 2, Safety-Related Maintenance Requirements.

Informational Note No. 5: See IEEE 1584, *Guide for Performing Arc Flash Calculations*, for more information regarding arc flash hazards for three-phase systems.

(A) Documentation. The results of the arc flash risk assessment shall be documented.

(B) Arc Flash Boundary.

(1) The arc flash boundary shall be the distance at which the incident energy equals 5 J/cm^2 (1.2 cal/cm^2).

Informational Note: For information on estimating the arc flash boundary, see Informative Annex D.

(2) The arc flash boundary shall be permitted to be determined by Table 130.7(C)(15)(A)(b) or Table 130.7(C)(15)(B), when the requirements of these tables apply.

(C) Arc Flash PPE. One of the following methods shall be used for the selection of PPE. Either, but not both, methods shall be permitted to be used on the same piece of equipment. The results of an incident energy analysis to specify an arc flash PPE Category in Table 130.7(C)(16) shall not be permitted.

(1) Incident Energy Analysis Method. The incident energy exposure level shall be based on the working distance of the employee's face and chest areas from a prospective arc source for the specific task to be performed. Arc-rated clothing and other PPE shall be used by the employee based on the incident energy exposure associated with the specific task. Recognizing that incident energy increases as the distance from the arc flash decreases, additional PPE shall be used for any parts of the body that are closer than the distance at which the incident energy was determined.

Informational Note: For information on estimating the incident energy, see Informative Annex D. For information on selection of arc-rated clothing and other PPE, see Table H.3(b) in Informative Annex H.

(2) Arc Flash PPE Categories Method. The requirements of 130.7(C)(15) and 130.7(C)(16) shall apply when the arc flash PPE category method is used for the selection of arc flash PPE.

(D) Equipment Labeling. Electrical equipment such as switchboards, panelboards, industrial control panels, meter socket enclosures, and motor control centers that are in other than dwelling units and that are likely to require examination, adjustment, servicing, or maintenance while energized shall be field-marked with a label containing all the following information:

(1) Nominal system voltage
(2) Arc flash boundary
(3) At least one of the following:
   a. Available incident energy and the corresponding working distance, or the arc flash PPE category in Table 130.7(C)(15)(A)(b) or Table 130.7(C)(15)(B) for the equipment, but not both
   b. Minimum arc rating of clothing
   c. Site-specific level of PPE

Exception: Labels applied prior to September 30, 2011 are acceptable if they contain the available incident energy or required level of PPE.

The method of calculating and the data to support the information for the label shall be documented. Where the review of the arc flash hazard risk assessment identifies a change that renders the label inaccurate, the label shall be updated.

The owner of the electrical equipment shall be responsible for the documentation, installation, and maintenance of the field-marked label.

130.6 Other Precautions for Personnel Activities.

(A) Alertness.

(1) When Electrical Hazards Might Exist. Employees shall be instructed to be alert at all times when they are working within the limited approach boundary of energized electrical conductors or circuit parts operating at 50 volts or more and in work situations when electrical hazards might exist.

(2) When Impaired. Employees shall not be permitted to work within the limited approach boundary of energized electrical conductors or circuit parts operating at 50 volts or more, or where other electrical hazards exist, while their alertness is recognizably impaired due to illness, fatigue, or other reasons.

(3) Changes in Scope. Employees shall be instructed to be alert for changes in the job or task that may lead the person outside of the electrically safe work condition or expose the person to additional hazards that were not part of the original plan.

(B) Blind Reaching. Employees shall be instructed not to reach blindly into areas that might contain exposed energized electrical conductors or circuit parts where an electrical hazard exists.

(C) Illumination.

(1) General. Employees shall not enter spaces where electrical hazards exist unless illumination is provided that enables the employees to perform the work safely.

(2) Obstructed View of Work Area. Where lack of illumination or an obstruction precludes observation of the work to be performed, employees shall not perform any task within the limited approach boundary of energized electrical conductors or circuit parts operating at 50 volts or more or where an electrical hazard exists.

(D) Conductive Articles Being Worn. Conductive articles of jewelry and clothing (such as watchbands, bracelets, rings, key chains, necklaces, metalized aprons, cloth with conductive thread, metal headgear, or metal frame glasses) shall not be worn within the restricted approach boundary or where they present an electrical contact hazard with exposed energized electrical conductors or circuit parts.

(E) Conductive Materials, Tools, and Equipment Being Handled.

(1) General. Conductive materials, tools, and equipment that are in contact with any part of an employee's body shall be handled in a manner that prevents accidental contact with energized electrical conductors or circuit parts. Such materials and equipment shall include, but are not limited to, long conductive objects, such as ducts, pipes and tubes, conductive hose and rope, metal-lined rules and scales, steel tapes, pulling lines, metal scaffold parts, structural members, bull floats, and chains.

(2) Approach to Energized Electrical Conductors and Circuit Parts. Means shall be employed to ensure that conductive materials approach exposed energized electrical conductors or circuit parts no closer than that permitted by 130.2.

(F) Confined or Enclosed Work Spaces. When an employee works in a confined or enclosed space (such as a manhole or vault) that contains exposed energized electrical conductors or circuit parts operating at 50 volts or more, or where an electrical hazard exists, the employer shall provide, and the employee shall use, protective shields, protective barriers, or insulating materials as necessary to avoid inadvertent contact with these parts and the effects of the electrical hazards.

(G) Doors and Hinged Panels. Doors, hinged panels, and the like shall be secured to prevent their swinging into an employee and causing the employee to contact exposed energized electrical conductors or circuit parts operating at 50 volts or more or where an electrical hazard exists if movement of the door, hinged panel, and the like is likely to create a hazard.

(H) Clear Spaces. Working space required by other codes and standards shall not be used for storage. This space shall be kept clear to permit safe operation and maintenance of electrical equipment.

(I) Housekeeping Duties. Employees shall not perform housekeeping duties inside the limited approach boundary where there is a possibility of contact with energized electrical conductors or circuit parts, unless adequate safeguards (such as insulating equipment or barriers) are provided to prevent contact. Electrically conductive cleaning materials (including conductive solids such as steel wool, metalized cloth, and silicone carbide, as well as conductive liquid solutions) shall not be used inside the limited approach boundary unless procedures to prevent electrical contact are followed.

(J) Occasional Use of Flammable Materials. Where flammable materials are present only occasionally, electric equipment capable of igniting them shall not be permitted to be used, unless measures are taken to prevent hazardous conditions from developing. Such materials shall include, but are not limited to, flammable gases, vapors, or liquids; combustible dust; and ignitible fibers or flyings.

Informational Note: Electrical installation requirements for locations where flammable materials are present on a regular basis are contained in *NFPA 70, National Electrical Code*.

(K) Anticipating Failure. When there is evidence that electric equipment could fail and injure employees, the electric equipment shall be de-energized, unless the employer can demonstrate that de-energizing introduces additional hazards or increased risk or is infeasible because of equipment design or operational limitation. Until the equipment is de-energized or repaired, employees shall be protected from hazards associated with the impending failure of the equipment by suitable barricades and other alerting techniques necessary for safety of the employees.

Informational Note: See 130.7(E) for alerting techniques.

(L) Routine Opening and Closing of Circuits. Load-rated switches, circuit breakers, or other devices specifically designed as disconnecting means shall be used for the opening, reversing, or closing of circuits under load conditions. Cable connectors not of the load-break type, fuses, terminal lugs, and cable splice connections shall not be permitted to be used for such purposes, except in an emergency.

(M) Reclosing Circuits After Protective Device Operation. After a circuit is de-energized by the automatic operation of a circuit protective device, the circuit shall not be manually reenergized until it has been determined that the equipment and circuit can be safely energized. The repetitive manual reclosing of circuit breakers or reenergizing circuits through replaced fuses shall be prohibited. When it is determined from the design of the circuit and the overcurrent devices involved that the automatic operation of a device was caused by an overload rather than a fault condition, examination of the circuit or connected equipment shall not be required before the circuit is reenergized.

(N) Safety Interlocks. Only qualified persons following the requirements for working inside the restricted approach boundary as covered by 130.4(C) shall be permitted to defeat or bypass an electrical safety interlock over which the person has sole control, and then only temporarily while the qualified person is working on the equipment. The safety interlock system shall be returned to its operable condition when the work is completed.

130.7 Personal and Other Protective Equipment.

(A) General. Employees working in areas where electrical hazards are present shall be provided with, and shall use, protective equipment that is designed and constructed for the specific part of the body to be protected and for the work to be performed.

Informational Note No. 1: The PPE requirements of 130.7 are intended to protect a person from arc flash and shock hazards. While some situations could result in burns to the skin, even with the protection selected, burn injury should be reduced and survivable. Due to the explosive effect of some arc events, physical trauma injuries could occur. The

PPE requirements of 130.7 do not address protection against physical trauma other than exposure to the thermal effects of an arc flash.

Informational Note No. 2: It is the collective experience of the Technical Committee on Electrical Safety in the Workplace that normal operation of enclosed electrical equipment, operating at 600 volts or less, that has been properly installed and maintained by qualified persons is not likely to expose the employee to an electrical hazard.

Informational Note No. 3: When incident energy exceeds 40 cal/cm^2 at the working distance, greater emphasis may be necessary with respect to de-energizing when exposed to electrical hazards.

(B) Care of Equipment. Protective equipment shall be maintained in a safe, reliable condition. The protective equipment shall be visually inspected before each use. Protective equipment shall be stored in a manner to prevent damage from physically damaging conditions and from moisture, dust, or other deteriorating agents.

Informational Note: Specific requirements for periodic testing of electrical protective equipment are given in 130.7(C)(14) and 130.7(F).

(C) Personal Protective Equipment (PPE).

(1) General. When an employee is working within the restricted approach boundary, the worker shall wear PPE in accordance with 130.4. When an employee is working within the arc flash boundary, he or she shall wear protective clothing and other PPE in accordance with 130.5. All parts of the body inside the arc flash boundary shall be protected.

(2) Movement and Visibility. When arc-rated clothing is worn to protect an employee, it shall cover all ignitible clothing and shall allow for movement and visibility.

(3) Head, Face, Neck, and Chin (Head Area) Protection. Employees shall wear nonconductive head protection wherever there is a danger of head injury from electric shock or burns due to contact with energized electrical conductors or circuit parts or from flying objects resulting from electrical explosion. Employees shall wear nonconductive protective equipment for the face, neck, and chin whenever there is a danger of injury from exposure to electric arcs or flashes or from flying objects resulting from electrical explosion. If employees use hairnets or beard nets, or both, these items must be arc rated.

Informational Note: See 130.7(C)(10)(b) and (c) for arc flash protective requirements.

(4) Eye Protection. Employees shall wear protective equipment for the eyes whenever there is danger of injury from electric arcs, flashes, or from flying objects resulting from electrical explosion.

(5) Hearing Protection. Employees shall wear hearing protection whenever working within the arc flash boundary.

(6) Body Protection. Employees shall wear arc-rated clothing wherever there is possible exposure to an electric arc flash above the threshold incident energy level for a second degree burn [5 J/cm^2 (1.2 cal/cm^2)].

(7) Hand and Arm Protection. Hand and arm protection shall be provided in accordance with 130.7(C)(7)(a), (b), and (c).

(a) Shock Protection. Employees shall wear rubber insulating gloves with leather protectors where there is a danger of hand injury from electric shock due to contact with energized electrical conductors or circuit parts. Employees shall wear rubber insulating gloves with leather protectors and rubber insulating sleeves where there is a danger of hand and arm injury from electric shock due to contact with energized electrical conductors or circuit parts. Rubber insulating gloves shall be rated for the voltage for which the gloves will be exposed.

Exception: Where it is necessary to use rubber insulating gloves without leather protectors, the requirements of ASTM F496, Standard Specification for In-Service Care of Insulating Gloves and Sleeves, shall be met.

• (b) Arc Flash Protection. Hand and arm protection shall be worn where there is possible exposure to arc flash burn. The apparel described in 130.7(C)(10)(d) shall be required for protection of hands from burns. Arm protection shall be accomplished by the apparel described in 130.7(C)(6).

(c) Maintenance and Use. Electrical protective equipment shall be maintained in a safe, reliable condition. Insulating equipment shall be inspected for damage before each day's use and immediately following any incident that can reasonably be suspected of having caused damage. Insulating gloves shall be given an air test, along with the inspection. Electrical protective equipment shall be subjected to periodic electrical tests. Test voltages and the maximum intervals between tests shall be in accordance with Table 130.7(C)(7)(c).

Informational Note: See OSHA 1910.137 and ASTM F496, *Standard Specification for In-Service Care of Insulating Gloves and Sleeves.*

(8) Foot Protection. Where insulated footwear is used as protection against step and touch potential, dielectric footwear shall be required. Insulated soles shall not be used as primary electrical protection.

Informational Note: Electrical hazard footwear meeting ASTM F2413, *Standard Specification for Performance Requirements for Protective (Safety) Toe Cap Footwear*, can provide a secondary source of electric shock protection under dry conditions.

Table 130.7(C)(7)(c) Rubber Insulating Equipment, Maximum Test Intervals

Rubber Insulating Equipment	When to Test	Governing Standard for Test Voltage*
Blankets	Before first issue; every 12 months thereafter†	ASTM F479
Covers	If insulating value is suspect	ASTM F478
Gloves	Before first issue; every 6 months thereafter†	ASTM F496
Line hose	If insulating value is suspect	ASTM F478
Sleeves	Before first issue; every 12 months thereafter†	ASTM F496

*ASTM F478, *Standard Specification for In-Service Care of Insulating Line Hose and Covers*; ASTM F479, *Standard Specification for In-Service Care of Insulating Blankets*; ASTM F496, *Standard Specification for In-Service Care of Insulating Gloves and Sleeves.*

†If the insulating equipment has been electrically tested but not issued for service, it is not permitted to be placed into service unless it has been electrically tested within the previous 12 months.

(9) Factors in Selection of Protective Clothing. Clothing and equipment that provide worker protection from shock and arc flash hazards shall be used. If arc-rated clothing is required, it shall cover associated parts of the body as well as all flammable apparel while allowing movement and visibility.

Clothing and equipment required for the degree of exposure shall be permitted to be worn alone or integrated with flammable, nonmelting apparel. Garments that are not arc rated shall not be permitted to be used to increase the arc rating of a garment or of a clothing system.

Informational Note: Protective clothing includes shirts, pants, coveralls, jackets, and parkas worn routinely by workers who, under normal working conditions, are exposed to momentary electric arc and related thermal hazards. Arc-rated rainwear worn in inclement weather is included in this category of clothing.

(a) Layering. Nonmelting, flammable fiber garments shall be permitted to be used as underlayers in conjunction with arc-rated garments in a layered system. If nonmelting, flammable fiber garments are used as underlayers, the system arc rating shall be sufficient to prevent breakopen of the innermost arc-rated layer at the expected arc exposure incident energy level to prevent ignition of flammable underlayers. Garments that are not arc rated shall not be permitted to be used to increase the arc rating of a garment or of a clothing system.

Informational Note: A typical layering system might include cotton underwear, a cotton shirt and trouser, and an arc-rated coverall. Specific tasks might call for additional arc-rated layers to achieve the required protection level.

(b) Outer Layers. Garments worn as outer layers over arc-rated clothing, such as jackets or rainwear, shall also be made from arc-rated material.

(c) Underlayers. Meltable fibers such as acetate, nylon, polyester, polypropylene, and spandex shall not be permitted in fabric underlayers (underwear) next to the skin.

Exception: An incidental amount of elastic used on nonmelting fabric underwear or socks shall be permitted.

Informational Note No. 1: Arc-rated garments (e.g., shirts, trousers, and coveralls) worn as underlayers that neither ignite nor melt and drip in the course of an exposure to electric arc and related thermal hazards generally provide a higher system arc rating than nonmelting, flammable fiber underlayers.

Informational Note No. 2: Arc-rated underwear or undergarments used as underlayers generally provide a higher system arc rating than nonmelting, flammable fiber underwear or undergarments used as underlayers.

(d) Coverage. Clothing shall cover potentially exposed areas as completely as possible. Shirt and coverall sleeves shall be fastened at the wrists, shirts shall be tucked into pants, and shirts, coveralls, and jackets shall be closed at the neck.

(e) Fit. Tight-fitting clothing shall be avoided. Loose-fitting clothing provides additional thermal insulation because of air spaces. Arc-rated apparel shall fit properly such that it does not interfere with the work task.

(f) Interference. The garment selected shall result in the least interference with the task but still provide the necessary protection. The work method, location, and task could influence the protective equipment selected.

(10) Arc Flash Protective Equipment.

(a) Arc Flash Suits. Arc flash suit design shall permit easy and rapid removal by the wearer. The entire arc flash suit, including the hood's face shield, shall have an arc rating that is suitable for the arc flash exposure. When exterior air is supplied into the hood, the air hoses and pump housing shall be either covered by arc-rated materials or constructed of nonmelting and nonflammable materials.

Paragraph 130.7(C)(10)(b)(1) was revised by a Tentative Interim Amendment (TIA). See page 1.

(b) Head Protection.

(1) An arc-rated balaclava shall be used with an arc-rated face shield when the back of the head is within the arc flash boundary. An arc-rated hood shall be permitted to be used instead of an arc-rated face shield and balaclava.
(2) An arc-rated hood shall be used when the anticipated incident energy exposure exceeds 12 cal/cm^2.

(c) Face Protection. Face shields shall have an arc rating suitable for the arc flash exposure. Face shields with a wrap-around guarding to protect the face, chin, forehead, ears, and neck area shall be used. Face shields without an arc rating shall not be used. Eye protection (safety glasses or goggles) shall always be worn under face shields or hoods.

Informational Note: Face shields made with energy-absorbing formulations that can provide higher levels of protection from the radiant energy of an arc flash are available, but these shields are tinted and can reduce visual acuity and color perception. Additional illumination of the task area might be necessary when these types of arc-protective face shields are used.

(d) Hand Protection.

(1) Heavy-duty leather gloves or arc-rated gloves shall be worn where required for arc flash protection.

Informational Note: Heavy-duty leather gloves are made entirely of leather with minimum thickness of 0.03 in. (0.7 mm) and are unlined or lined with nonflammable, nonmelting fabrics. Heavy-duty leather gloves meeting this requirement have been shown to have ATPV values in excess of 10 cal/cm^2.

(2) Where insulating rubber gloves are used for shock protection, leather protectors shall be worn over the rubber gloves.

Informational Note: The leather protectors worn over rubber insulating gloves provide additional arc flash protection for the hands for arc flash protection exposure.

(e) Foot Protection. Heavy-duty leather footwear provide some arc flash protection to the feet and shall be used in all exposures greater than 4 cal/cm^2.

(11) Clothing Material Characteristics. Arc-rated clothing shall meet the requirements described in 130.7(C)(14) and 130.7(C)(12).

Informational Note No. 1: Arc-rated materials, such as flame-retardant-treated cotton, meta-aramid, para-aramid, and poly-benzimidazole (PBI) fibers, provide thermal protection. These materials can ignite but will not continue to burn after the ignition source is removed. Arc-rated fabrics can reduce burn injuries during an arc flash exposure by providing a thermal barrier between the arc flash and the wearer.

Informational Note No. 2: Non–arc-rated cotton, polyester-cotton blends, nylon, nylon-cotton blends, silk, rayon, and wool fabrics are flammable. Fabrics, zipper tapes, and findings made of these materials can ignite and continue to burn on the body, resulting in serious burn injuries.

Informational Note No. 3: Rayon is a cellulose-based (wood pulp) synthetic fiber that is a flammable but nonmelting material.

Clothing consisting of fabrics, zipper tapes, and findings made from flammable synthetic materials that melt at temperatures below 315°C (600°F), such as acetate, acrylic, nylon, polyester, polyethylene, polypropylene, and spandex, either alone or in blends, shall not be used.

Informational Note: These materials melt as a result of arc flash exposure conditions, form intimate contact with the skin, and aggravate the burn injury.

Exception: Fiber blends that contain materials that melt, such as acetate, acrylic, nylon, polyester, polyethylene, polypropylene, and spandex, shall be permitted if such blends in fabrics meet the requirements of ASTM F1506, Standard Performance Specification for Flame Resistant and Arc Rated Textile Materials for Wearing Apparel for Use by Electrical Workers Exposed to Momentary Electric Arc and Related Thermal Hazards, and if such blends in fabrics do not exhibit evidence of a melting and sticking hazard during arc testing according to ASTM F1959/F1959M, Standard Test Method for Determining the Arc Rating of Materials for Clothing.

(12) Clothing and Other Apparel Not Permitted. Clothing and other apparel (such as hard hat liners and hair nets) made from materials that do not meet the requirements of 130.7(C)(11) regarding melting or made from materials that do not meet the flammability requirements shall not be permitted to be worn.

Informational Note: Some flame-resistant fabrics, such as non-flame-resistant modacrylic and nondurable flame-retardant treatments of cotton, are not recommended for industrial electrical or utility applications.

Exception No. 1: Nonmelting, flammable (non–arc-rated) materials shall be permitted to be used as underlayers to arc-rated clothing, as described in 130.7(C)(11).

Exception No. 2: Where the work to be performed inside the arc flash boundary exposes the worker to multiple hazards, such as airborne contaminants, and the risk assessment identifies that the level of protection is adequate to address the arc flash hazard, non–arc-rated PPE shall be permitted.

(13) Care and Maintenance of Arc-Rated Clothing and Arc-Rated Arc Flash Suits.

(a) Inspection. Arc-rated apparel shall be inspected before each use. Work clothing or arc flash suits that are contaminated or damaged to the extent that their protective qualities are impaired shall not be used. Protective items that become contaminated with grease, oil, or flammable liquids or combustible materials shall not be used.

(b) Manufacturer's Instructions. The garment manufacturer's instructions for care and maintenance of arc-rated apparel shall be followed.

(c) Storage. Arc-rated apparel shall be stored in a manner that prevents physical damage; damage from moisture, dust, or other deteriorating agents; or contamination from flammable or combustible materials.

(d) Cleaning, Repairing, and Affixing Items. When arc-rated clothing is cleaned, manufacturer's instructions shall be followed to avoid loss of protection. When arc-rated clothing is repaired, the same arc-rated materials used to manufacture the arc-rated clothing shall be used to provide repairs.

Informational Note No. 1: Additional guidance is provided in ASTM F1506, *Standard Performance Specification for Flame Resistant and Arc Rated Textile Materials for Wearing Apparel for Use by Electrical Workers Exposed to Momentary Electric Arc and Related Thermal Hazards*, when trim, name tags, logos, or any combination thereof are affixed to arc-rated clothing.

Informational Note No. 2: Additional guidance is provided in ASTM F1449, *Standard Guide for Industrial Laundering of Flame, Thermal, and Arc Resistant Clothing*, and ASTM F2757, *Standard Guide for Home Laundering Care and Maintenance of Flame, Thermal, and Arc Resistant Clothing*.

(14) Standards for Personal Protective Equipment (PPE). PPE shall conform to the standards listed in Table 130.7(C)(14).

Informational Note: Non–arc-rated or flammable fabrics are not covered by any of the standards in Table 130.7(C)(14). See 130.7(C)(11) and 130.7(C)(12).

(15) Selection of Personal Protective Equipment (PPE) When Required for Various Tasks.

(A Alternating Current (ac) Equipment. When selected in lieu of the incident energy analysis of 130.5(B)(1), Table 130.7(C)(15)(A)(a) shall be used to identify when arc flash PPE is required. When arc flash PPE is required, Table 130.7(C)(15)(A)(b) shall be used to determine the arc flash PPE category. The estimated maximum available short-circuit current, maximum fault-clearing times, and minimum working distances for various ac equipment types or classifications are listed in Table 130.7(C)(15)(A)(b). An incident energy analysis shall be required in accordance with 130.5 for the following:

(1) Tasks not listed in Table 130.7(C)(15)(A)(a)

(2) Power systems with greater than the estimated maximum available short-circuit current

(3) Power systems with longer than the maximum fault clearing times

(4) Tasks with less than the minimum working distance

(B) Direct Current (dc) Equipment. When selected in lieu of the incident energy analysis of 130.5(C)(1), Table 130.7(C)(15)(A)(a) shall be used to identify when arc flash PPE is required. When arc flash PPE is required, Table 130.7(C)(15)(B) shall be used to determine the arc flash PPE category. The estimated maximum available short circuit current, maximum arc duration and working distances for dc equipment are listed in 130.7(C)(15)(B). An incident energy analysis shall be required in accordance with 130.5 for the following:

(1) Tasks not listed in Table 130.7(C)(15)(A)(a)

(2) Power systems with greater than the estimated maximum available short circuit current

(3) Power systems with longer than the maximum fault clearing times

(4) Tasks with less than the minimum working distance

Informational Note No. 1: The arc flash PPE category, work tasks, and protective equipment provided in Table 130.7(C)(15)(A)(a), Table 130.7(C)(15)(A)(b), and Table 130.7(C)(15)(B) were identified and selected, based on the collective experience of the NFPA 70E Technical Committee. The arc flash PPE category of the protective clothing and equipment is generally based on determination of the estimated exposure level.

Informational Note No. 2: The collective experience of the NFPA 70E Technical Committee is that, in most cases, closed doors do not provide enough protection to eliminate the need for PPE in situations in which the state of the equipment is known to readily change (e.g., doors open or closed, rack in or rack out).

Informational Note No. 3: The premise used by the NFPA 70E Technical Committee in developing the criteria discussed in Informational Note No. 1 and Informational Note No. 2 is considered to be reasonable, based on the consensus judgment of the committee.

(16) Protective Clothing and Personal Protective Equipment (PPE). Once the arc flash PPE category has been identified from Table 130.7(C)(15)(A)(b) or Table 130.7(C)(15)(B), Table 130.7(C)(16) shall be used to determine the required PPE for the task. Table 130.7(C)(16) lists the requirements for PPE based on arc flash PPE categories 1 through 4. This clothing and equipment shall be used when working within the arc flash boundary.

Informational Note No. 1: See Informative Annex H for a suggested simplified approach to ensure adequate PPE for electrical workers within facilities with large and diverse electrical systems.

Informational Note No. 2: The PPE requirements of this section are intended to protect a person from arc flash hazards. While some situations could result in burns to the skin, even with the protection described in Table 130.7(C)(16), burn injury should be reduced and survivable. Due to the explosive effect of some arc events, physical trauma injuries could occur. The PPE requirements of this section do not address protection against physical trauma other than exposure to the thermal effects of an arc flash.

Informational Note No. 3: The arc rating for a particular clothing system can be obtained from the arc-rated clothing manufacturer.

Table 130.7(C)(14) Standards on Protective Equipment

Subject	Document Title	Document Number
Apparel-Arc Rated	Standard Performance Specification for Flame Resistant and Arc Rated Textile Materials for Wearing Apparel for Use by Electrical Workers Exposed to Momentary Electric Arc and Related Thermal Hazards	ASTM F1506
	Standard Guide for Industrial Laundering of Flame, Thermal, and Arc Resistant Clothing	ASTM F1449
	Standard Guide for Home Laundering Care and Maintenance of Flame, Thermal and Arc Resistant Clothing	ASTM F2757
Aprons-Insulating	Standard Specification for Electrically Insulating Aprons	ASTM F2677
Eye and Face Protection-General	Practice for Occupational and Educational Eye and Face Protection	ANSI Z87.1
Face-Arc Rated	Standard Test Method for Determining the Arc Rating and Standard Specification for Eye or Face Protective Products	ASTM F2178
Fall Protection	Standard Specification for Personal Climbing Equipment	ASTM F887
Footwear-Dielectric Specification	Standard Specification for Dielectric Footwear	ASTM F1117
Footwear-Dielectric Test Method	Standard Test Method for Determining Dielectric Strength of Dielectric Footwear	ASTM F1116
Footwear-Standard Performance Specification	Standard Specification for Performance Requirements for Protective (Safety) Toe Cap Footwear	ASTM F2413
Footwear-Standard Test Method	Standard Test Methods for Foot Protections	ASTM F2412
Gloves-Leather Protectors	Standard Specification for Leather Protectors for Rubber Insulating Gloves and Mittens	ASTM F696
Gloves-Rubber Insulating	Standard Specification for Rubber Insulating Gloves	ASTM D120
Gloves and Sleeves –In-Service Care	Standard Specification for In-Service Care of Insulating Gloves and Sleeves	ASTM F496
Head Protection-Hard Hats	Requirements for Protective Headwear for Industrial Workers	ANSI Z89.1
Rainwear-Arc Rated	Standard Specification for Arc and Flame Resistant Rainwear	ASTM F1891
Rubber Protective Products-Visual Inspection	Standard Guide for Visual Inspection of Electrical Protective Rubber Products	ASTM F1236
Sleeves-Insulating	Standard Specification for Rubber Insulating Sleeves	ASTM D1051

Table 130.7(C)(15)(A)(a) Arc Flash Hazard Identification for Alternating Current (ac) and Direct Current (dc) Systems

Task	Equipment Condition*	Arc Flash PPE Required
Reading a panel meter while operating a meter switch	Any	No
Normal operation of a circuit breaker (CB), switch, contactor, or starter	All of the following: The equipment is properly installed The equipment is properly maintained All equipment doors are closed and secured All equipment covers are in place and secured There is no evidence of impending failure	No
	One or more of the following: The equipment is not properly installed The equipment is not properly maintained Equipment doors are open or not secured Equipment covers are off or not secured There is evidence of impending failure	Yes
For ac systems: Work on energized electrical conductors and circuit parts, including voltage testing	Any	Yes
For dc systems: Work on energized electrical conductors and circuit parts of series-connected battery cells, including voltage testing	Any	Yes
Voltage testing on individual battery cells or individual multi-cell units	All of the following: The equipment is properly installed The equipment is properly maintained Covers for all other equipment are in place and secured There is no evidence of impending failure	No
	One or more of the following: The equipment is not properly installed The equipment is not properly maintained Equipment doors are open or not secured Equipment covers are off or not secured There is evidence of impending failure	Yes
Removal or installation of CBs or switches	Any	Yes
Removal or installation of covers for equipment such as wireways, junction boxes, and cable trays that does not expose bare energized electrical conductors and circuit parts	All of the following: The equipment is properly installed The equipment is properly maintained There is no evidence of impending failure	No
	Any of the following: The equipment is not properly installed The equipment is not properly maintained There is evidence of impending failure	Yes
Removal of bolted covers (to expose bare energized electrical conductors and circuit parts). For dc systems, this includes bolted covers, such as battery terminal covers.	Any	Yes

(continues)

Table 130.7(C)(15)(A)(a) ***Continued***

Task	Equipment Condition*	Arc Flash PPE Required
Removal of battery intercell connector covers	All of the following: The equipment is properly installed. The equipment is properly maintained Covers for all other equipment are in place and secured There is no evidence of impending failure	No
	One or more of the following: The equipment is not properly installed The equipment is not properly maintained Equipment doors are open or not secured Equipment covers are off or not secured There is evidence of impending failure	Yes
Opening hinged door(s) or cover(s) (to expose bare energized electrical conductors and circuit parts)	Any	Yes
Perform infrared thermography and other noncontact inspections outside the restricted approach boundary. This activity does not include opening of doors or covers.	Any	No
Application of temporary protective grounding equipment after voltage test	Any	Yes
Work on control circuits with exposed energized electrical conductors and circuit parts, 120 volts or below without any other exposed energized equipment over 120 V including opening of hinged covers to gain access	Any	No
Work on control circuits with exposed energized electrical conductors and circuit parts, greater than 120 V	Any	Yes
Insertion or removal of individual starter buckets from motor control center (MCC)	Any	Yes
Insertion or removal (racking) of CBs or starters from cubicles, doors open or closed	Any	Yes
Insertion or removal of plug-in devices into or from busways	Any	Yes
Insulated cable examination with no manipulation of cable	Any	No
Insulated cable examination with manipulation of cable	Any	Yes
Work on exposed energized electrical conductors and circuit parts of equipment directly supplied by a panelboard or motor control center	Any	Yes
Insertion and removal of revenue meters (kW-hour, at primary voltage and current)	Any	Yes
For dc systems, insertion or removal of individual cells or multi-cell units of a battery system in an enclosure	Any	Yes
For dc systems, insertion or removal of individual cells or multi-cell units of a battery system in an open rack	Any	No

Table 130.7(C)(15)(A)(a) ***Continued***

Task	Equipment Condition*	Arc Flash PPE Required
For dc systems, maintenance on a single cell of a battery system or multi-cell units in an open rack	Any	No
For dc systems, work on exposed energized electrical conductors and circuit parts of utilization equipment directly supplied by a dc source	Any	Yes
Arc-resistant switchgear Type 1 or 2 (for clearing times of <0.5 sec with a prospective fault current not to exceed the arc-resistant rating of the equipment) and metal enclosed interrupter switchgear, fused or unfused of arc resistant type construction, tested in accordance with IEEE C37.20.7: •Insertion or removal (racking) of CBs from cubicles •Insertion or removal (racking) of ground and test device •Insertion or removal (racking) of voltage transformers on or off the bus	All of the following: The equipment is properly installed The equipment is properly maintained All equipment doors are closed and secured All equipment covers are in place and secured There is no evidence of impending failure	No
	One or more of the following: The equipment is not properly installed The equipment is not properly maintained Equipment doors are open or not secured Equipment covers are off or not secured There is evidence of impending failure	Yes
Opening voltage transformer or control power transformer compartments	Any	Yes
Outdoor disconnect switch operation (hookstick operated) at 1 kV through 15 kV	Any	Yes
Outdoor disconnect switch operation (gang-operated, from grade) at 1 kV through 15 kV	Any	Yes

Note: Hazard identification is one component of risk assessment. Risk assessment involves a determination of the likelihood of occurrence of an incident, resulting from a hazard that could cause injury or damage to health. The assessment of the likelihood of occurrence contained in this table does not cover every possible condition or situation. Where this table indicates that arc flash PPE is not required, an arc flash is not likely to occur.

*The phrase *properly installed*, as used in this table, means that the equipment is installed in accordance with applicable industry codes and standards and the manufacturer's recommendations. The phrase *properly maintained*, as used in this table, means that the equipment has been maintained in accordance with the manufacturer's recommendations and applicable industry codes and standards. The phrase *evidence of impending failure*, as used in this table, means that there is evidence of arcing, overheating, loose or bound equipment parts, visible damage, deterioration, or other damage.

Table 130.7(C)(15)(A)(b) Arc-Flash Hazard PPE Categories for Alternating Current (ac) Systems

Equipment	Arc Flash PPE Category	Arc-Flash Boundary
Panelboards or other equipment rated 240 V and below Parameters: Maximum of 25 kA short-circuit current available; maximum of 0.03 sec (2 cycles) fault clearing time; working distance 455 mm (18 in.)	1	485 mm (19 in.)
Panelboards or other equipment rated >240 V and up to 600 V Parameters: Maximum of 25 kA short-circuit current available; maximum of 0.03 sec (2 cycles) fault clearing time; working distance 455 mm (18 in.)	2	900 mm (3 ft)
600-V class motor control centers (MCCs) Parameters: Maximum of 65 kA short-circuit current available; maximum of 0.03 sec (2 cycles) fault clearing time; working distance 455 mm (18 in.)	2	1.5 m (5 ft)
600-V class motor control centers (MCCs) Parameters: Maximum of 42 kA short-circuit current available; maximum of 0.33 sec (20 cycles) fault clearing time; working distance 455 mm (18 in.)	4	4.3 m (14 ft)
600-V class switchgear (with power circuit breakers or fused switches) and 600 V class switchboards Parameters: Maximum of 35 kA short-circuit current available; maximum of up to 0.5 sec (30 cycles) fault clearing time; working distance 455 mm (18 in.)	4	6 m (20 ft)
Other 600-V class (277 V through 600 V, nominal) equipment Parameters: Maximum of 65 kA short circuit current available; maximum of 0.03 sec (2 cycles) fault clearing time; working distance 455 mm (18 in.)	2	1.5 m (5 ft)
NEMA E2 (fused contactor) motor starters, 2.3 kV through 7.2 kV Parameters: Maximum of 35 kA short-circuit current available; maximum of up to 0.24 sec (15 cycles) fault clearing time; working distance 910 mm (36 in.)	4	12 m (40 ft)
Metal-clad switchgear, 1 kV through 15 kV Parameters: Maximum of 35 kA short-circuit current available; maximum of up to 0.24 sec (15 cycles) fault clearing time; working distance 910 mm (36 in.)	4	12 m (40 ft)
Arc-resistant switchgear Type 1 or 2 [for clearing times of < 0.5 sec (30 cycles) with a perspective fault current not to exceed the arc-resistant rating of the equipment], and metal-enclosed interrupter switchgear, fused or unfused of arc-resistant-type construction, tested in accordance with IEEE C37.20.7, 1 kV through 15 kV	N/A (doors closed)	N/A (doors closed)
Parameters: Maximum of 35 kA short-circuit current available; maximum of up to 0.24 sec (15 cycles) fault clearing time; working distance 910 mm (36 in.)	4 (doors open)	12 m (40 ft)
Other equipment 1 kV through 15 kV Parameters: Maximum of 35 kA short-circuit current available; maximum of up to 0.24 sec (15 cycles) fault clearing time; working distance 910 mm (36 in.)	4	12 m (40 ft)

Note: For equipment rated 600 volts and below, and protected by upstream current-limiting fuses or current-limiting circuit breakers sized at 200 amperes or less, the arc flash PPE category can be reduced by one number but not below arc flash PPE category 1.

Table 130.7(C)(15)(B) Arc-Flash Hazard PPE Categories for Direct Current (dc) Systems

Equipment	Arc Flash PPE Category	Arc-Flash Boundary
Storage batteries, dc switchboards, and other dc supply sources 100 V > Voltage < 250 V Parameters: Voltage: 250 V Maximum arc duration and working distance: 2 sec @ 455 mm (18 in.)		
Short-circuit current < 4 kA	1	900 mm (3 ft)
4 kA ≤ short-circuit current < 7 kA	2	1.2 m (4 ft)
7 kA ≤ short-circuit current < 15 kA	3	1.8 m (6 ft)
Storage batteries, dc switchboards, and other dc supply sources 250 V ≤ Voltage ≤ 600 V Parameters: Voltage: 600 V Maximum arc duration and working distance: 2 sec @ 455 mm (18 in.)		
Short-circuit current <1.5 kA	1	900 mm (3 ft)
1.5 kA ≤ short-circuit current < 3 kA	2	1.2 m (4 ft)
3 kA ≤ short-circuit current < 7 kA	3	1.8 m (6 ft.)
7 kA ≤ short-circuit current < 10 kA	4	2.5 m (8 ft)

Note: Apparel that can be expected to be exposed to electrolyte must meet both of the following conditions:
(1) Be evaluated for electrolyte protection in accordance with ASTM F1296, *Standard Guide for Evaluating Chemical Protective Clothing*
(2) Be arc-rated in accordance with ASTM F1891, *Standard Specification for Arc Rated and Flame Resistant Rainwear*, or equivalent

Informational Note No. 1: "Short-circuit current," as used in this table, is determined from the dc power system maximum available short-circuit, including the effects of cables and any other impedances in the circuit. Power system modeling is the best method to determine the available short-circuit current at the point of the arc. Battery cell short-circuit current can be obtained from the battery manufacturer. See Informative Annex D.5 for the basis for table values and alternative methods to determine dc incident energy. Methods should be used with good engineering judgment.

Informational Note No. 2: The methods for estimating the dc arc flash incident energy that were used to determine the categories for this table are based on open-air incident energy calculations. Open-air calculations were used because many battery systems and other dc process systems are in open areas or rooms. If the specific task is within in an enclosure, it would be prudent to consider additional PPE protection beyond the value shown in this table. Research with ac arc flash has shown a multiplier of as much as 3x for arc-in-a-box [508 mm (20 in.) cube] versus open air. Engineering judgment is required when reviewing the specific conditions of the equipment and task to be performed, including the dimensions of the enclosure and the working distance involved.

Table 130.7(C)(16) Personal Protective Equipment (PPE)

PPE Category	PPE
1	**Arc-Rated Clothing, Minimum Arc Rating of 4 cal/cm²** (see Note 1) Arc-rated long-sleeve shirt and pants or arc-rated coverall Arc-rated face shield (see Note 2) or arc flash suit hood Arc-rated jacket, parka, rainwear, or hard hat liner (AN) **Protective Equipment** Hard hat Safety glasses or safety goggles (SR) Hearing protection (ear canal inserts) Heavy duty leather gloves (see Note 3) Leather footwear (AN)
2	**Arc-Rated Clothing, Minimum Arc Rating of 8 cal/cm²** (see Note 1) Arc-rated long-sleeve shirt and pants or arc-rated coverall Arc-rated flash suit hood or arc-rated face shield (see Note 2) and arc-rated balaclava Arc-rated jacket, parka, rainwear, or hard hat liner (AN) **Protective Equipment** Hard hat Safety glasses or safety goggles (SR) Hearing protection (ear canal inserts) Heavy duty leather gloves (see Note 3) Leather footwear
3	**Arc-Rated Clothing Selected so That the System Arc Rating Meets the Required Minimum Arc Rating of 25 cal/cm²** (see Note 1) Arc-rated long-sleeve shirt (AR) Arc-rated pants (AR) Arc-rated coverall (AR) Arc-rated arc flash suit jacket (AR) Arc-rated arc flash suit pants (AR) Arc-rated arc flash suit hood Arc-rated gloves (see Note 3) Arc-rated jacket, parka, rainwear, or hard hat liner (AN) **Protective Equipment** Hard hat Safety glasses or safety goggles (SR) Hearing protection (ear canal inserts) Leather footwear
4	**Arc-Rated Clothing Selected so That the System Arc Rating Meets the Required Minimum Arc Rating of 40 cal/cm²** (see Note 1) Arc-rated long-sleeve shirt (AR) Arc-rated pants (AR) Arc-rated coverall (AR) Arc-rated arc flash suit jacket (AR) Arc-rated arc flash suit pants (AR) Arc-rated arc flash suit hood Arc-rated gloves (see Note 3) Arc-rated jacket, parka, rainwear, or hard hat liner (AN) **Protective Equipment** Hard hat Safety glasses or safety goggles (SR) Hearing protection (ear canal inserts) Leather footwear

AN: as needed (optional). AR: as required. SR: selection required.
Notes:
(1) *Arc rating* is defined in Article 100.
(2) Face shields are to have wrap-around guarding to protect not only the face but also the forehead, ears, and neck, or, alternatively, an arc-rated arc flash suit hood is required to be worn.
(3) If rubber insulating gloves with leather protectors are used, additional leather or arc-rated gloves are not required. The combination of rubber insulating gloves with leather protectors satisfies the arc flash protection requirement.

(D) Other Protective Equipment.

(1) Insulated Tools and Equipment. Employees shall use insulated tools or handling equipment, or both, when working inside the restricted approach boundary of exposed energized electrical conductors or circuit parts where tools or handling equipment might make accidental contact. Insulated tools shall be protected from damage to the insulating material.

Informational Note: See 130.4(B), Shock Protection Boundaries.

(a) Requirements for Insulated Tools. The following requirements shall apply to insulated tools:

(1) Insulated tools shall be rated for the voltages on which they are used.
(2) Insulated tools shall be designed and constructed for the environment to which they are exposed and the manner in which they are used.
(3) Insulated tools and equipment shall be inspected prior to each use. The inspection shall look for damage to the insulation or damage that can limit the tool from performing its intended function or could increase the potential for an incident (e.g., damaged tip on a screwdriver).

(b) Fuse or Fuseholder Handling Equipment. Fuse or fuseholder handling equipment, insulated for the circuit voltage, shall be used to remove or install a fuse if the fuse terminals are energized.

(c) Ropes and Handlines. Ropes and handlines used within the limited approach boundary of exposed energized electrical conductors or circuit parts operating at 50 volts or more, or used where an electrical hazard exists, shall be nonconductive.

(d) Fiberglass-Reinforced Plastic Rods. Fiberglass-reinforced plastic rod and tube used for live-line tools shall meet the requirements of applicable portions of electrical codes and standards dealing with electrical installation requirements.

Informational Note: For further information concerning electrical codes and standards dealing with installation requirements, refer to ASTM F 711, *Standard Specification for Fiberglass-Reinforced Plastic (FRP) Rod and Tube Used in Live Line Tools.*

(e) Portable Ladders. Portable ladders shall have nonconductive side rails if they are used where an employee or ladder could contact exposed energized electrical conductors or circuit parts operating at 50 volts or more or where an electrical hazard exists. Nonconductive ladders shall meet the requirements of ANSI standards for ladders listed in Table 130.7(F).

(f) Protective Shields. Protective shields, protective barriers, or insulating materials shall be used to protect each employee from shock, burns, or other electrically related injuries while an employee is working within the limited approach boundary of energized conductors or circuit parts that might be accidentally contacted or where dangerous electric heating or arcing might occur. When normally enclosed energized conductors or circuit parts are exposed for maintenance or repair, they shall be guarded to protect unqualified persons from contact with the energized conductors or circuit parts.

(g) Rubber Insulating Equipment. Rubber insulating equipment used for protection from accidental contact with energized conductors or circuit parts shall meet the requirements of the ASTM standards listed in Table 130.7(F).

(h) Voltage-Rated Plastic Guard Equipment. Plastic guard equipment for protection of employees from accidental contact with energized conductors or circuit parts, or for protection of employees or energized equipment or material from contact with ground, shall meet the requirements of the ASTM standards listed in Table 130.7(F).

(i) Physical or Mechanical Barriers. Physical or mechanical (field-fabricated) barriers shall be installed no closer than the restricted approach boundary distance given in Table 130.4(D)(a) and Table 130.4(D)(b). While the barrier is being installed, the restricted approach boundary distance specified in Table 130.4(D)(a) and Table 130.4(D)(b) shall be maintained, or the energized conductors or circuit parts shall be placed in an electrically safe work condition.

(E) Alerting Techniques.

(1) Safety Signs and Tags. Safety signs, safety symbols, or accident prevention tags shall be used where necessary to warn employees about electrical hazards that might endanger them. Such signs and tags shall meet the requirements of ANSI Z535, *Series of Standards for Safety Signs and Tags*, given in Table 130.7(F).

Informational Note: Safety signs, tags, and barricades used to identify energized "look-alike" equipment can be employed as an additional preventive measure.

(2) Barricades. Barricades shall be used in conjunction with safety signs where it is necessary to prevent or limit employee access to work areas containing energized conductors or circuit parts. Conductive barricades shall not be used where it might increase the likelihood of exposure to an electrical hazard. Barricades shall be placed no closer than the limited approach boundary given in Table 130.4(D)(a) and Table 130.4(D)(b). Where the arc flash boundary is greater than the limited approach boundary, barricades shall not be placed closer than the arc flash boundary.

(3) Attendants. If signs and barricades do not provide sufficient warning and protection from electrical hazards, an attendant shall be stationed to warn and protect employees. The primary duty and responsibility of an attendant providing manual signaling and alerting shall be to keep unqualified employees outside a work area where the unqualified employee might be exposed to electrical hazards. An attendant shall remain in the area as long as there is a potential for employees to be exposed to the electrical hazards.

(4) Look-Alike Equipment. Where work performed on equipment that is de-energized and placed in an electrically safe condition exists in a work area with other energized equipment that is similar in size, shape, and construction, one of the alerting methods in 130.7(E)(1), (2), or (3) shall be employed to prevent the employee from entering look-alike equipment.

(F) Standards for Other Protective Equipment. Other protective equipment required in 130.7(D) shall conform to the standards given in Table 130.7(F).

130.8 Work Within the Limited Approach Boundary or Arc Flash Boundary of Overhead Lines.

(A) Uninsulated and Energized. Where work is performed in locations containing uninsulated energized overhead lines that are not guarded or isolated, precautions shall be taken to prevent employees from contacting such lines directly with any unguarded parts of their body or indirectly through conductive materials, tools, or equipment. Where the work to be performed is such that contact with uninsulated energized overhead lines is possible, the lines shall be de-energized and visibly grounded at the point of work or suitably guarded.

Table 130.7(F) Standards on Other Protective Equipment

Subject	Document	Document Number
Arc Protective Blankets	Standard Test Method for Determining the Protective Performance of an Arc Protective Blanket for Electric Arc Hazards	ASTM F2676
Blankets	Standard Specification for Rubber Insulating Blankets	ASTM D1048
Blankets — In-service Care	Standard Specification for In-Service Care of Insulating Blankets	ASTM F479
Covers	Standard Specification for Rubber Covers	ASTM D1049
Fiberglass Rods — Live Line Tools	Standard Specification for Fiberglass-Reinforced Plastic (FRP) Rod and Tube Used in Live Line Tools	ASTM F711
Insulated Hand Tools	Standard Specification for Insulated and Insulating Hand Tools	ASTM F1505
Ladders	American National Standard for Ladders — Wood — Safety Requirements	ANSI/ASC A14.1
	American National Standard for Ladders — Fixed — Safety Requirements	ANSI/ASC A14.3
	American National Standard Safety Requirements for Job Made Ladders	ANSI ASC A14.4
	American National Standard for Ladders-Portable Reinforced-Safety Requirements	ANSI ASC A14.5
Line Hose	Standard Specification for Rubber Insulating Line Hoses	ASTM D1050
Line Hose and Covers — In-service Care	Standard Specification for In-Service Care of Insulating Line Hose and Covers	ASTM F478
Plastic Guard	Standard Test Methods and Specifications for Electrically Insulating Plastic Guard Equipment for Protection of Workers	ASTM F712
Sheeting	Standard Specification for PVC Insulating Sheeting	ASTM F1742
	Standard Specification for Rubber Insulating Sheeting	ASTM F2320
Safety Signs and Tags	Series of Standards for Safety Signs and Tags	ANSI Z535
Shield Performance on Live Line Tool	Standard Test Method for Determining the Protective Performance of a Shield Attached on Live Line Tools or on Racking Rods for Electric Arc Hazards	ASTM F2522
Temporary Protective Grounds — In-service Testing	Standard Specification for In-Service Test Methods for Temporary Grounding Jumper Assemblies Used on De-energized Electric Power Lines and Equipment	ASTM F2249
Temporary Protective Grounds — Test Specification	Standard Specification for Temporary Protective Grounds to Be Used on De-energized Electric Power Lines and Equipment	ASTM F855

(B) Determination of Insulation Rating. A qualified person shall determine if the overhead electrical lines are insulated for the lines' operating voltage.

(C) De-energizing or Guarding. If the lines are to be de-energized, arrangements shall be made with the person or organization that operates or controls the lines to de-energize them and visibly ground them at the point of work. If arrangements are made to use protective measures, such as guarding, isolating, or insulation, these precautions shall prevent each employee from contacting such lines directly with any part of his or her body or indirectly through conductive materials, tools, or equipment.

(D) Employer and Employee Responsibility. The employer and employee shall be responsible for ensuring that guards or protective measures are satisfactory for the conditions. Employees shall comply with established work methods and the use of protective equipment.

(E) Approach Distances for Unqualified Persons. When unqualified persons are working on the ground or in an elevated position near overhead lines, the location shall be such that the employee and the longest conductive object the employee might contact do not come closer to any unguarded, energized overhead power line than the limited approach boundary in Table 130.4(D)(a), column 2 or Table 130.4(D)(b), column 2.

Informational Note: Objects that are not insulated for the voltage involved should be considered to be conductive.

(F) Vehicular and Mechanical Equipment.

(1) Elevated Equipment. Where any vehicle or mechanical equipment structure will be elevated near energized overhead lines, it shall be operated so that the limited approach boundary distance of Table 130.4(D)(a), column 2 or Table 130.4(D)(b), column 2, is maintained. However, under any of the following conditions, the clearances shall be permitted to be reduced:

(1) If the vehicle is in transit with its structure lowered, the limited approach boundary to overhead lines in Table 130.4(D)(a), column 2 or Table 130.4(D)(b), column 2, shall be permitted to be reduced by 1.83 m (6 ft). If insulated barriers, rated for the voltages involved, are installed and they are not part of an attachment to the vehicle, the clearance shall be permitted to be reduced to the design working dimensions of the insulating barrier.

(2) If the equipment is an aerial lift insulated for the voltage involved, and if the work is performed by a qualified person, the clearance (between the uninsulated portion of the aerial lift and the power line) shall be permitted to be reduced to the restricted approach boundary given in Table 130.4(D)(a), column 4 or Table 130.4(D)(b), column 4.

(2) Equipment Contact. Employees standing on the ground shall not contact the vehicle or mechanical equipment or any of its attachments unless either of the following conditions apply:

(1) The employee is using protective equipment rated for the voltage.

(2) The equipment is located so that no uninsulated part of its structure (that portion of the structure that provides a conductive path to employees on the ground) can come closer to the line than permitted in 130.8(F)(1).

(3) Equipment Grounding. If any vehicle or mechanical equipment capable of having parts of its structure elevated near energized overhead lines is intentionally grounded, employees working on the ground near the point of grounding shall not stand at the grounding location whenever there is a possibility of overhead line contact. Additional precautions, such as the use of barricades, dielectric overshoe footwear, or insulation, shall be taken to protect employees from hazardous ground potentials (step and touch potential).

Informational Note: Upon contact of the elevated structure with the energized lines, hazardous ground potentials can develop within a few feet or more outward from the grounded point.

130.9 Underground Electrical Lines and Equipment. Before excavation starts where there exists a reasonable possibility of contacting electrical lines or equipment, the employer shall take the necessary steps to contact the appropriate owners or authorities to identify and mark the location of the electrical lines or equipment. When it has been determined that a reasonable possibility of contacting electrical lines or equipment exists, appropriate safe work practices and PPE shall be used during the excavation.

130.10 Cutting or Drilling. Before cutting or drilling into equipment, floors, walls, or structural elements where a likelihood of contacting energized electrical lines or parts exists, the employer shall perform a risk assessment to:

(1) Identify and mark the location of conductors, cables, raceways, or equipment

(2) Create an electrically safe work condition

(3) Identify safe work practices and PPE to be used

Chapter 2 Safety-Related Maintenance Requirements

ARTICLE 200 Introduction

200.1 Scope. Chapter 2 addresses the requirements that follow.

(1) Chapter 2 covers practical safety-related maintenance requirements for electrical equipment and installations in workplaces as included in 90.2. These requirements identify only that maintenance directly associated with employee safety.

(2) Chapter 2 does not prescribe specific maintenance methods or testing procedures. It is left to the employer to choose from the various maintenance methods available to satisfy the requirements of Chapter 2.

(3) For the purpose of Chapter 2, maintenance shall be defined as preserving or restoring the condition of electrical equipment and installations, or parts of either, for the safety of employees who work where exposed to electrical hazards. Repair or replacement of individual portions or parts of equipment shall be permitted without requiring modification or replacement of other portions or parts that are in a safe condition.

Informational Note: Refer to NFPA 70B, *Recommended Practice for Electrical Equipment Maintenance*; ANSI/NETA MTS, *Standard for Maintenance Testing Specifications for Electrical Power Distribution Equipment and Systems*; and IEEE 3007.2, *IEEE Recommended Practice for the Maintenance of Industrial and Commercial Power Systems*, for guidance on maintenance frequency, methods, and tests.

ARTICLE 205 General Maintenance Requirements

205.1 Qualified Persons. Employees who perform maintenance on electrical equipment and installations shall be qualified persons as required in Chapter 1 and shall be trained in, and familiar with, the specific maintenance procedures and tests required.

205.2 Single-Line Diagram. A single-line diagram, where provided for the electrical system, shall be maintained in a legible condition and shall be kept current.

205.3 General Maintenance Requirements. Electrical equipment shall be maintained in accordance with manufacturers' instructions or industry consensus standards to reduce the risk associated with failure. The equipment owner or the owner's designated representative shall be responsible for maintenance of the electrical equipment and documentation.

Informational Note: Common industry practice is to apply test or calibration decals to equipment to indicate the test or calibration date and overall condition of equipment that has been tested and maintained in the field. These decals provide the employee immediate indication of last maintenance date and if the tested device or system was found acceptable on the date of test. This local information can assist the employee in the assessment of overall electrical equipment maintenance status.

205.4 Overcurrent Protective Devices. Overcurrent protective devices shall be maintained in accordance with the manufacturers' instructions or industry consensus standards. Maintenance, tests, and inspections shall be documented.

205.5 Spaces About Electrical Equipment. All working space and clearances required by electrical codes and standards shall be maintained.

Informational Note: For further information concerning spaces about electrical equipment, see Article 110, Parts II and III, of *NFPA 70, National Electrical Code.*

205.6 Grounding and Bonding. Equipment, raceway, cable tray, and enclosure bonding and grounding shall be maintained to ensure electrical continuity.

205.7 Guarding of Energized Conductors and Circuit Parts. Enclosures shall be maintained to guard against accidental contact with energized conductors and circuit parts and other electrical hazards. Covers and doors shall be in place with all associated fasteners and latches secured.

205.8 Safety Equipment. Locks, interlocks, and other safety equipment shall be maintained in proper working condition to accomplish the control purpose.

205.9 Clear Spaces. Access to working space and escape passages shall be kept clear and unobstructed.

205.10 Identification of Components. Identification of components, where required, and safety-related instructions (operating or maintenance), if posted, shall be securely attached and maintained in legible condition.

205.11 Warning Signs. Warning signs, where required, shall be visible, securely attached, and maintained in legible condition.

205.12 Identification of Circuits. Circuit or voltage identification shall be securely affixed and maintained in updated and legible condition.

205.13 Single and Multiple Conductors and Cables. Electrical cables and single and multiple conductors shall be maintained free of damage, shorts, and ground that would expose employees to an electrical hazard.

205.14 Flexible Cords and Cables. Flexible cords and cables shall be maintained to preserve insulation integrity.

(1) Damaged Cords and Cables. Cords and cables shall not have worn, frayed, or damaged areas that would expose employees to an electrical hazard.

(2) Strain Relief. Strain relief of cords and cables shall be maintained to prevent pull from being transmitted directly to joints or terminals.

(3) Repair and Replacement. Cords and cord caps for portable electrical equipment shall be repaired and replaced by qualified personnel and checked for proper polarity, grounding, and continuity prior to returning to service.

205.15 Overhead Line Clearances. For overhead electric lines under the employer's control, grade elevation shall be maintained to preserve no less than the minimum designed vertical and horizontal clearances necessary to minimize risk of unintentional contact.

ARTICLE 210 Substations, Switchgear Assemblies, Switchboards, Panelboards, Motor Control Centers, and Disconnect Switches

210.1 Enclosures. Enclosures shall be kept free of material that would expose employees to an electrical hazard.

210.2 Area Enclosures. Fences, physical protection, enclosures, or other protective means, where required to guard against unauthorized access or accidental contact with exposed energized conductors and circuit parts, shall be maintained.

210.3 Conductors. Current-carrying conductors (buses, switches, disconnects, joints, and terminations) and bracing shall be maintained to perform as follows:

(1) Conduct rated current without overheating

(2) Withstand available fault current

210.4 Insulation Integrity. Insulation integrity shall be maintained to support the voltage impressed.

210.5 Protective Devices. Protective devices shall be maintained to adequately withstand or interrupt available fault current.

Informational Note: Improper or inadequate maintenance can result in increased opening time of the overcurrent protective device, thus increasing the incident energy.

ARTICLE 215 Premises Wiring

215.1 Covers for Wiring System Components. Covers for wiring system components shall be in place with all associated hardware, and there shall be no unprotected openings.

215.2 Open Wiring Protection. Open wiring protection, such as location or barriers, shall be maintained to prevent accidental contact.

215.3 Raceways and Cable Trays. Raceways and cable trays shall be maintained to provide physical protection and support for conductors.

ARTICLE 220 Controller Equipment

220.1 Scope. This article shall apply to controllers, including electrical equipment that governs the starting, stopping, direction of motion, acceleration, speed, and protection of rotating equipment and other power utilization apparatus in the workplace.

220.2 Protection and Control Circuitry. Protection and control circuitry used to guard against accidental contact with energized conductors and circuit parts and to prevent other electrical or mechanical hazards shall be maintained.

ARTICLE 225 Fuses and Circuit Breakers

225.1 Fuses. Fuses shall be maintained free of breaks or cracks in fuse cases, ferrules, and insulators. Fuse clips shall be maintained to provide adequate contact with fuses. Fuseholders for current-limiting fuses shall not be modified to allow the insertion of fuses that are not current-limiting. Non-current limiting fuses shall not be modified to allow their insertion into current-limiting fuseholders.

225.2 Molded-Case Circuit Breakers. Molded-case circuit breakers shall be maintained free of cracks in cases and cracked or broken operating handles.

225.3 Circuit Breaker Testing After Electrical Faults. Circuit breakers that interrupt faults approaching their interrupting ratings shall be inspected and tested in accordance with the manufacturer's instructions.

ARTICLE 230 Rotating Equipment

230.1 Terminal Boxes. Terminal chambers, enclosures, and terminal boxes shall be maintained to guard against accidental contact with energized conductors and circuit parts and other electrical hazards.

230.2 Guards, Barriers, and Access Plates. Guards, barriers, and access plates shall be maintained to prevent employees from contacting moving or energized parts.

ARTICLE 235 Hazardous (Classified) Locations

235.1 Scope. This article covers maintenance requirements in those areas identified as hazardous (classified) locations.

Informational Note No. 1: These locations need special types of equipment and installation to ensure safe performance under conditions of proper use and maintenance. It is important that inspection authorities and users exercise more than ordinary care with regard to installation and maintenance. The maintenance for specific equipment and materials is covered elsewhere in Chapter 2 and is applicable to hazardous (classified) locations. Other maintenance will ensure that the form of construction and of installation that makes the equipment and materials suitable for the particular location are not compromised.

Informational Note No. 2: The maintenance needed for specific hazardous (classified) locations depends on the classification of the specific location. The design principles and equipment characteristics, for example, use of positive pressure ventilation, explosionproof, nonincendive, intrinsically safe, and purged and pressurized equipment, that were applied in the installation to meet the requirements of the area classification must also be known. With this information, the employer and the inspection authority are able to determine whether the installation as maintained has retained the condition necessary for a safe workplace.

235.2 Maintenance Requirements for Hazardous (Classified) Locations. Equipment and installations in these locations shall be maintained such that the following criteria are met:

(1) No energized parts are exposed.

Exception to (1): Intrinsically safe and nonincendive circuits.

(2) There are no breaks in conduit systems, fittings, or enclosures from damage, corrosion, or other causes.

(3) All bonding jumpers are securely fastened and intact.

(4) All fittings, boxes, and enclosures with bolted covers have all bolts installed and bolted tight.

(5) All threaded conduit are wrenchtight and enclosure covers are tightened in accordance with the manufacturer's instructions.

(6) There are no open entries into fittings, boxes, or enclosures that would compromise the protection characteristics.

(7) All close-up plugs, breathers, seals, and drains are securely in place.

(8) Marking of luminaires (lighting fixtures) for maximum lamp wattage and temperature rating is legible and not exceeded.

(9) Required markings are secure and legible.

ARTICLE 240 Batteries and Battery Rooms

240.1 Ventilation. Ventilation systems, forced or natural, shall be maintained to prevent buildup of explosive mixtures. This maintenance shall include a functional test of any associated detection and alarm systems.

240.2 Eye and Body Wash Apparatus. Eye and body wash apparatus shall be maintained in operable condition.

ARTICLE 245 Portable Electric Tools and Equipment

245.1 Maintenance Requirements for Portable Electric Tools and Equipment. Attachment plugs, receptacles, cover plates, and cord connectors shall be maintained such that the following criteria are met:

(1) There are no breaks, damage, or cracks exposing energized conductors and circuit parts.
(2) There are no missing cover plates.
(3) Terminations have no stray strands or loose terminals.
(4) There are no missing, loose, altered, or damaged blades, pins, or contacts.
(5) Polarity is correct.

ARTICLE 250 Personal Safety and Protective Equipment

250.1 Maintenance Requirements for Personal Safety and Protective Equipment. Personal safety and protective equipment such as the following shall be maintained in a safe working condition:

(1) Grounding equipment
(2) Hot sticks
(3) Rubber gloves, sleeves, and leather protectors
(4) Test instruments
(5) Blanket and similar insulating equipment
(6) Insulating mats and similar insulating equipment
(7) Protective barriers
(8) External circuit breaker rack-out devices
(9) Portable lighting units
(10) Temporary protective grounding equipment
(11) Dielectric footwear
(12) Protective clothing
(13) Bypass jumpers
(14) Insulated and insulating hand tools

250.2 Inspection and Testing of Protective Equipment and Protective Tools.

(A) Visual. Safety and protective equipment and protective tools shall be visually inspected for damage and defects before initial use and at intervals thereafter, as service conditions require, but in no case shall the interval exceed 1 year, unless specified otherwise by the respective ASTM standards.

(B) Testing. The insulation of protective equipment and protective tools, such as items specified in 250.1(1) through 250.1 (14), that is used as primary protection from shock hazards and requires an insulation system to ensure protection of personnel, shall be verified by the appropriate test and visual inspection to ascertain that insulating capability has been retained before initial use, and at intervals thereafter, as service conditions and applicable standards and instructions require, but in no case shall the interval exceed 3 years.

250.3 Safety Grounding Equipment.

(A) Visual. Personal protective ground cable sets shall be inspected for cuts in the protective sheath and damage to the conductors. Clamps and connector strain relief devices shall be checked for tightness. These inspections shall be made at intervals thereafter as service conditions require, but in no case shall the interval exceed 1 year.

(B) Testing. Prior to being returned to service, temporary protective grounding equipment that has been repaired or modified shall be tested.

Informational Note: Guidance for inspecting and testing safety grounds is provided in ASTM F2249, *Standard Specification for In-Service Test Methods for Temporary Grounding Jumper Assemblies Used on De-Energized Electric Power Lines and Equipment.*

(C) Grounding and Testing Devices. Grounding and testing devices shall be stored in a clean and dry area. Grounding and testing devices shall be properly inspected and tested before each use.

Informational Note: Guidance for testing of grounding and testing devices is provided in Section 9.5 of IEEE C37.20.6, *Standard for 4.76 kV to 38 kV-Rated Ground and Test Devices Used in Enclosures.*

250.4 Test Instruments. Test instruments and associated test leads used to verify the absence or presence of voltage shall be maintained to assure functional integrity. The maintenance program shall include functional verification as described in 110.4(A)(5).

Chapter 3 Safety Requirements for Special Equipment

ARTICLE 300 Introduction

300.1 Scope. Chapter 3 covers special electrical equipment in the workplace and modifies the general requirements of Chapter 1.

300.2 Responsibility. The employer shall provide safety-related work practices and employee training. The employee shall follow those work practices.

300.3 Organization. Chapter 3 of this standard is divided into articles. Article 300 applies generally. Article 310 applies to electrolytic cells. Article 320 applies to batteries and battery rooms. Article 330 applies to lasers. Article 340 applies to power electronic equipment. Article 350 applies to research and development (R&D) laboratories.

ARTICLE 310 Safety-Related Work Practices for Electrolytic Cells

310.1 Scope. The requirements of this article shall apply to the electrical safety-related work practices used in the types of electrolytic cell areas.

Informational Note No. 1: See Informative Annex L for a typical application of safeguards in the cell line working zone.

Informational Note No. 2: For further information about electrolytic cells, see *NFPA 70, National Electrical Code*, Article 668.

310.2 Definitions. For the purposes of this article, the definitions that follow shall apply.

Battery Effect. A voltage that exists on the cell line after the power supply is disconnected.

Informational Note: Electrolytic cells can exhibit characteristics similar to an electrical storage battery and a shock hazard could exist after the power supply is disconnected from the cell line.

Safeguarding. Safeguards for personnel include the consistent administrative enforcement of safe work practices. Safeguards include training in safe work practices, cell line design, safety equipment, PPE, operating procedures, and work checklists.

310.3 Safety Training.

(A) General. The training requirements of this chapter shall apply to employees exposed to electrical hazards in the cell line working zone defined in 110.2 and shall supplement or modify the requirements of 120.1, 130.2, 130.3, and 130.8.

(B) Training Requirements. Employees shall be trained to understand the specific electrical hazards associated with electrical energy on the cell line. Employees shall be trained in safety-related work practices and procedural requirements to provide protection from the electrical hazards associated with their respective job or task assignment.

310.4 Employee Training.

(A) Qualified Persons.

(1) Training. Qualified persons shall be trained and knowledgeable in the operation of cell line working zone equipment and specific work methods and shall be trained to avoid the electrical hazards that are present. Such persons shall be familiar with the proper use of precautionary techniques and PPE. Training for a qualified person shall include the following:

(1) Skills and techniques to avoid a shock hazard:
   a. Between energized surfaces, which might include temporarily insulating or guarding parts to permit the employee to work on energized parts
   b. Between energized surfaces and grounded equipment, other grounded objects, or the earth itself, that might include temporarily insulating or guarding parts to permit the employee to work on energized parts

(2) Method of determining the cell line working zone area boundaries

(2) Qualified Persons. Qualified persons shall be permitted to work within the cell line working zone.

(B) Unqualified Persons.

(1) Training. Unqualified persons shall be trained to identify electrical hazards to which they could be exposed and the proper methods of avoiding the hazards.

(2) In Cell Line Working Zone. When there is a need for an unqualified person to enter the cell line working zone to perform a specific task, that person shall be advised of the electrical hazards by the designated qualified person in charge to ensure that the unqualified person is safeguarded.

310.5 Safeguarding of Employees in the Cell Line Working Zone.

(A) General. Operation and maintenance of electrolytic cell lines might require contact by employees with exposed energized surfaces such as buses, electrolytic cells, and their attachments. The approach distances referred to in Table 130.4(D)(a) and Table 130.4(D)(b) shall not apply to work performed by qualified persons in the cell line working zone. Safeguards such as safety-related work practices and other safeguards shall be used to protect employees from injury while working in the cell line working zone. These safeguards shall be consistent with the nature and extent of the related electrical hazards. Safeguards might be different for energized cell lines and de-energized cell lines. Hazardous battery effect voltages shall be dissipated to consider a cell line de-energized.

Informational Note No. 1: Exposed energized surfaces might not present an electrical hazard. Electrical hazards are related to current flow through the body, causing shock and arc flash burns and arc blasts. Shock is a function of many factors, including resistance through the body and the skin, return paths, paths in parallel with the body, and system voltages. Arc flash burns and arc blasts are a function of the current available at the point involved and the time of arc exposure.

Informational Note No. 2: A cell line or group of cell lines operated as a unit for the production of a particular metal, gas, or chemical compound might differ from other cell lines producing the same product because of variations in the particular raw materials used, output capacity, use of proprietary methods or process practices, or other modifying factors. Detailed standard electrical safety-related work practice requirements could become overly restrictive without accomplishing the stated purpose of Chapter 1.

(B) Signs. Permanent signs shall clearly designate electrolytic cell areas.

(C) Electrical Arc Flash Hazard Analysis. The requirements of 130.5, Arc Flash Risk Assessment, shall not apply to electrolytic cell line work zones.

(1) Arc Flash Risk Assessment. Each task performed in the electrolytic cell line working zone shall be analyzed for the likelihood of arc flash injury. If there is a likelihood of personal injury, appropriate measures shall be taken to protect persons exposed to the arc flash hazards, including one or more of the following:

(1) Providing appropriate PPE *[see 310.5(D)(2)]* to prevent injury from the arc flash hazard
(2) Altering work procedures to reduce the likelihood of occurrence of an arc flash incident
(3) Scheduling the task so that work can be performed when the cell line is de-energized

(2) Routine Tasks. Arc flash risk assessment shall be done for all routine tasks performed in the cell line work zone. The results of the arc flash risk assessment shall be used in training employees in job procedures that minimize the possibility of arc flash hazards. The training shall be included in the requirements of 310.3.

(3) Nonroutine Tasks. Before a nonroutine task is performed in the cell line working zone, an arc flash risk assessment shall be done. If an arc flash hazard is a possibility during nonroutine work, appropriate instructions shall be given to employees involved on how to minimize the risk associated with arc flash.

(4) Arc Flash Hazards. If the likelihood of occurrence of an arc flash hazard exists for either routine or nonroutine tasks, employees shall use appropriate safeguards.

(D) Safeguards. Safeguards shall include one or a combination of the following means.

(1) Insulation. Insulation shall be suitable for the specific conditions, and its components shall be permitted to include glass, porcelain, epoxy coating, rubber, fiberglass, and plastic and, when dry, such materials as concrete, tile, brick, and wood. Insulation shall be permitted to be applied to energized or grounded surfaces.

(2) Personal Protective Equipment (PPE). PPE shall provide protection from electrical hazards. PPE shall include one or more of the following, as determined by authorized management:

(1) Footwear for wet service
(2) Gloves for wet service
(3) Sleeves for wet service
(4) Footwear for dry service
(5) Gloves for dry service
(6) Sleeves for dry service
(7) Electrically insulated head protection
(8) Protective clothing
(9) Eye protection with nonconductive frames
(10) Face shield (polycarbonate or similar nonmelting type)

a. Standards for PPE. Personal and other protective equipment shall be appropriate for conditions, as determined by authorized management, and shall not be required to meet the equipment standards in 130.7(C)(14) through 130.7(F) and in Table 130.7(C)(14) and Table 130.7(F).
b. Testing of PPE. PPE shall be verified with regularity and by methods that are consistent with the exposure of employees to electrical hazards.

(3) Barriers. Barriers shall be devices that prevent contact with energized or grounded surfaces that could present an electrical hazard.

(4) Voltage Equalization. Voltage equalization shall be permitted by bonding a conductive surface to an electrically energized surface, either directly or through a resistance, so that there is insufficient voltage to create an electrical hazard.

(5) Isolation. Isolation shall be the placement of equipment or items in locations such that employees are unable to simultaneously contact exposed conductive surfaces that could present an electrical hazard.

(6) Safe Work Practices. Employees shall be trained in safe work practices. The training shall include why the work practices in a cell line working zone are different from similar work situations in other areas of the plant. Employees shall comply with established safe work practices and the safe use of protective equipment.

(a) Attitude Awareness. Safe work practice training shall include attitude awareness instruction. Simultaneous contact with energized parts and ground can cause serious electrical shock. Of special importance is the need to be aware of body position where contact may be made with energized parts of the electrolytic cell line and grounded surfaces.

(b) Bypassing of Safety Equipment. Safe work practice training shall include techniques to prevent bypassing the protection of safety equipment. Clothing may bypass protective equipment if the clothing is wet. Trouser legs should be kept at appropriate length, and shirt sleeves should be a good fit so as not to drape while reaching. Jewelry and other metal accessories that may bypass protective equipment shall not be worn while working in the cell line working zone.

(7) Tools. Tools and other devices used in the energized cell line work zone shall be selected to prevent bridging between surfaces at hazardous potential difference.

Informational Note: Tools and other devices of magnetic material could be difficult to handle in the area of energized cells due to their strong dc magnetic fields.

(8) Portable Cutout-Type Switches. Portable cell cutout switches that are connected shall be considered as energized and as an extension of the cell line working zone. Appropriate procedures shall be used to ensure proper cutout switch connection and operation.

(9) Cranes and Hoists. Cranes and hoists shall meet the requirements of 668.32 of *NFPA 70, National Electrical Code*. Insulation required for safeguarding employees, such as insulated crane hooks, shall be periodically tested.

(10) Attachments. Attachments that extend the cell line electrical hazards beyond the cell line working zone shall use one or more of the following:

(1) Temporary or permanent extension of the cell line working zone
(2) Barriers
(3) Insulating breaks
(4) Isolation

(11) Pacemakers and Metallic Implants. Employees with implanted pacemakers, ferromagnetic medical devices, or other electronic devices vital to life shall not be permitted in cell areas unless written permission is obtained from the employee's physician.

Informational Note: The American Conference of Government Industrial Hygienists (ACGIH) recommends that persons with implanted pacemakers should not be exposed to magnetic flux densities above 10 gauss.

(12) Testing. Equipment safeguards for employee protection shall be tested to ensure they are in a safe working condition.

310.6 Portable Tools and Equipment.

Informational Note: The order of preference for the energy source for portable hand-held equipment is considered to be as follows:

(1) Battery power
(2) Pneumatic power
(3) Portable generator
(4) Nongrounded-type receptacle connected to an ungrounded source

(A) Portable Electrical Equipment. The grounding requirements of 110.4(B)(2) shall not be permitted within an energized cell line working zone. Portable electrical equipment shall meet the requirements of 668.20 of *NFPA 70, National Electrical Code*. Power supplies for portable electric equipment shall meet the requirements of 668.21 of *NFPA 70*.

(B) Auxiliary Nonelectric Connections. Auxiliary nonelectric connections such as air, water, and gas hoses shall meet the requirements of 668.31 of *NFPA 70, National Electrical Code*. Pneumatic-powered tools and equipment shall be supplied with nonconductive air hoses in the cell line working zone.

(C) Welding Machines. Welding machine frames shall be considered at cell potential when within the cell line working zone. Safety-related work practices shall require that the cell line not be grounded through the welding machine or its power supply. Welding machines located outside the cell line working zone shall be barricaded to prevent employees from touching the welding machine and ground simultaneously where the welding cables are in the cell line working zone.

(D) Portable Test Equipment. Test equipment in the cell line working zone shall be suitable for use in areas of large magnetic fields and orientation.

Informational Note: Test equipment that is not suitable for use in such magnetic fields could result in an incorrect response. When such test equipment is removed from the cell line working zone, its performance might return to normal, giving the false impression that the results were correct.

ARTICLE 320 Safety Requirements Related to Batteries and Battery Rooms

320.1 Scope. This article covers electrical safety requirements for the practical safeguarding of employees while working with exposed stationary storage batteries that exceed 50 volts, nominal.

Informational Note: For additional information on best practices for safely working on stationary batteries, see the following documents:

(1) NFPA 1, *Fire Code*, Chapter 52, Stationary Storage Battery Systems, 2015
(2) *NFPA 70, National Electrical Code*, Article 480, Storage Batteries, 2014
(3) IEEE 450, *IEEE Recommended Practice for Maintenance, Testing, and Replacement of Vented Lead-Acid Batteries for Stationary Applications*, 2010
(4) IEEE 937, *Recommended Practice for Installation and Maintenance of Lead-Acid Batteries for Photovoltaic Systems*, 2007
(5) IEEE 1106, *IEEE Recommended Practice for Installation, Maintenance, Testing, and Replacement of Vented Nickel-Cadmium Batteries for Stationary Applications*, 2005 (R 2011)
(6) IEEE 1184, *IEEE Guide for Batteries for Uninterruptible Power Supply Systems*, 2006
(7) IEEE 1188, *IEEE Recommended Practice for Maintenance, Testing, and Replacement of Valve-Regulated Lead-Acid (VRLA) Batteries for Stationary Applications*, 2005 (R 2010)
(8) IEEE 1657, *Recommended Practice for Personnel Qualifications for Installation and Maintenance of Stationary Batteries,* 2009
(9) OSHA 29 CFR 1910.305(j)(7), "Storage batteries"
(10) OSHA 29 CFR 1926.441, "Batteries and battery charging"
(11) DHHS (NIOSH) Publication No. 94-110, *Applications Manual for the Revised NIOSH Lifting Equation,* 1994

320.2 Definitions. For the purposes of this article definitions that follow shall apply.

Authorized Personnel. The person in charge of the premises, or other persons appointed or selected by the person in charge of the premises who performs certain duties associated with stationary storage batteries.

Battery. A system consisting of two or more electrochemical cells connected in series or parallel and capable of storing electrical energy received and that can give it back by reconversion.

Battery Room. A room specifically intended for the installation of batteries that have no other protective enclosure.

Cell. The basic electrochemical unit, characterized by an anode and a cathode used to receive, store, and deliver electrical energy.

Electrolyte. A solid, liquid, or aqueous immobilized liquid medium that provides the ion transport mechanism between the positive and negative electrodes of a cell.

Nominal Voltage. The value assigned to a cell or battery of a given voltage class for the purpose of convenient designation; the operating voltage of the cell or system may vary above or below this value.

Pilot Cell. One or more cells chosen to represent the operating parameters of the entire battery (sometimes called "temperature reference" cell).

Prospective Short-Circuit Current. The highest level of fault current that could theoretically occur at a point on a circuit. This is the fault current that can flow in the event of a zero impedance short circuit and if no protection devices operate.

Valve-Regulated Lead Acid (VRLA) Cell. A lead-acid cell that is sealed with the exception of a valve that opens to the atmosphere when the internal pressure in the cell exceeds atmospheric pressure by a pre-selected amount, and that provides a means for recombination of internally generated oxygen and the suppression of hydrogen gas evolution to limit water consumption.

Vented Cell. A type of cell in which the products of electrolysis and evaporation are allowed to escape freely into the atmosphere as they are generated. (Also called "flooded cell.")

320.3 Safety Procedures.

(A) General Safety Hazards.

(1) Battery Risk Assessment. Prior to any work on a battery system, a risk assessment shall be performed to identify the chemical, electrical shock, and arc flash hazards and assess the risks associated with the type of tasks to be performed.

(2) Battery Room or Enclosure Requirements.

(a) Personnel Access to Energized Batteries. Each battery room or battery enclosure shall be accessible only to authorized personnel.

(b) Illumination. Employees shall not enter spaces containing batteries unless illumination is provided that enables the employees to perform the work safely.

Informational Note: Battery terminals are normally exposed and pose possible shock hazard. Batteries are also installed in steps or tiers that can cause obstructions.

(3) Apparel. Personnel shall not wear electrically conductive objects such as jewelry while working on a battery system.

(4) Abnormal Battery Conditions. Instrumentation that provides alarms for early warning of abnormal conditions of battery operation, if present, shall be tested annually.

Informational Note: Battery monitoring systems typically include alarms for such conditions as overvoltage, undervoltage, overcurrent, ground fault, and overtemperature. The type of conditions monitored will vary depending upon the battery technology. One source of guidance on monitoring battery systems is IEEE 1491, *Guide for the Selection and Use of Battery Monitoring Equipment in Stationary Applications.*

(5) Warning Signs. The following warning signs or labels shall be posted in appropriate locations:

(1) Electrical hazard warnings indicating the shock hazard due to the battery voltage and the arc flash hazard due to the prospective short-circuit current, and the thermal hazard.

Informational Note No. 1: Because internal resistance, prospective short-circuit current, or both are not always provided on battery container labels or data sheets, and because many variables can be introduced into a battery layout, the battery manufacturer should be consulted for accurate data. Variables can include, but are not limited to, the following:

a. Series connections
b. Parallel connections
c. Charging methodology
d. Temperature
e. Charge status
f. Dc distribution cable size and length

Informational Note No. 2: See 130.5(D) for requirements for equipment labeling.

(2) Chemical hazard warnings, applicable to the worst case when multiple battery types are installed in the same space, indicating the following:

a. Potential presence of explosive gas (when applicable to the battery type)
b. Prohibition of open flame and smoking
c. Danger of chemical burns from the electrolyte (when applicable to the battery type)

(3) Notice for personnel to use and wear protective equipment and apparel appropriate to the hazard for the battery

(4) Notice prohibiting access to unauthorized personnel

(B) Electrolyte Hazards.

(1) Battery Activities That Include Handling of Liquid Electrolyte. The following protective equipment shall be available to employees performing any type of service on a battery with liquid electrolyte:

(1) Goggles and face shield appropriate for the electrical hazard and the chemical hazard
(2) Gloves and aprons appropriate for the chemical hazards
(3) Portable or stationary eye wash facilities within the work area that are capable of drenching or flushing of the eyes and body for the duration necessary to the hazard.

Informational Note: Guidelines for the use and maintenance of eye wash facilities for vented batteries in nontelecom environments can be found in ANSI/ISEA Z358.1, *American National Standard for Emergency Eye Wash and Shower Equipment.*

(2) Activities That Do Not Include Handling of Electrolyte. Employees performing any activity not involving the handling of electrolyte shall wear safety glasses.

• Informational Note: Battery maintenance activities usually do not involve handling electrolyte. Batteries with solid electrolyte (such as most lithium batteries) or immobilized electrolyte (such as valve-regulated lead acid batteries) present little or no electrolyte hazard. Most modern density meters expose a worker to a quantity of electrolyte too minute to be considered hazardous, if at all. Such work would not be considered handling electrolyte. However, if specific gravity readings are taken using a bulb hydrometer, the risk of exposure is higher — this could be considered to be handling electrolyte, and the requirements of 320.3(B)(1) would apply.

(C) Testing, Maintenance, and Operation.

• **(1) Direct-Current Ground-Fault Detection.** Ground-fault detection shall be based on the type of dc grounding systems utilized.

Informational Note: Not all battery systems have dc ground-fault detection systems. For personnel safety reasons, it is important to understand the grounding methodology being used and to determine the appropriate manner of detecting ground faults. If an unintended ground develops within the system (e.g., dirt and acid touching the battery rack), it can create a short circuit that could cause a fire. Commonly used dc grounding systems include, but are not limited to, the following:

(1) Type 1. An ungrounded dc system, in which neither pole of the battery is connected to ground. If an unintentional ground occurs at any place in the battery, an increased potential would exist, allowing fault current to flow between the opposite end of the battery and the ground. An ungrounded dc system is typically equipped with an alarm to indicate the presence of a ground fault.
(2) Type 2. A solidly grounded dc system, in which either the most positive or most negative pole of the battery is connected directly to ground. If an unintentional ground occurs, it introduces a path through which fault current can flow. A ground detection system is not typically used on this type of grounded system.

(3) Type 3. A resistance grounded dc system, which is a variation of a Type 1 system, in which the battery is connected to ground through a resistance. Detection of a change in the resistance typically enables activation of a ground-fault alarm. Introducing an unintentional ground at one point of the battery could be detected and alarmed. A second unintentional ground at a different point in the battery would create a path for short-circuit current to flow.
(4) Type 4. A solidly grounded dc system, either at the center point or at another point to suit the load system. If an unintentional ground occurs on either polarity, it introduces a path through which short circuit current can flow. A ground detection system is not typically used on this type of grounded system.

(2) Tools and Equipment.

(a) Tools and equipment for work on batteries shall be equipped with handles listed as insulated for the maximum working voltage.

(b) Battery terminals and all electrical conductors shall be kept clear of unintended contact with tools, test equipment, liquid containers, and other foreign objects.

(c) Nonsparking tools shall be required when the risk assessment required by 110.1(F) justifies their use.

(D) Cell Flame Arresters and Cell Ventilation. When present, battery cell ventilation openings shall be unobstructed. Cell flame arresters shall be inspected for proper installation and unobstructed ventilation and shall be replaced when necessary in accordance with the manufacturer's instructions.

ARTICLE 330 Safety-Related Work Practices for Use of Lasers

330.1 Scope. The requirements of this article shall apply to the use of lasers in the laboratory and the workshop.

330.2 Definitions. For the purposes of this article, the definitions that follow shall apply.

Fail Safe. The design consideration in which failure of a component does not create additional hazards or increased risk. In the failure mode, the system is rendered inoperative or nonhazardous.

Fail-Safe Safety Interlock. An interlock that in the failure mode does not defeat the purpose of the interlock; for example, an interlock that is positively driven into the off position as soon as a hinged cover begins to open, or before a detachable cover is removed, and that is positively held in the off position until the hinged cover is closed or the detachable cover is locked in the closed position.

Laser. Any device that can be made to produce or amplify electromagnetic radiation in the wavelength range from 100 nm to 1 mm primarily by the process of controlled stimulated emission.

Laser Energy Source. Any device intended for use in conjunction with a laser to supply energy for the excitation of electrons, ions, or molecules. General energy sources, such as electrical supply services or batteries, shall not be considered to constitute laser energy sources.

Laser Product. Any product or assembly of components that constitutes, incorporates, or is intended to incorporate a laser or laser system.

Laser Radiation. All electromagnetic radiation emitted by a laser product between 100 nm and 1 mm that is produced as a result of a controlled stimulated emission.

Laser System. A laser in combination with an appropriate laser energy source with or without additional incorporated components.

330.3 Safety Training.

(A) Personnel to Be Trained. Employers shall provide training for all operator and maintenance personnel.

(B) Scope of Training. The training shall include, but is not limited to, the following:

(1) Familiarization with laser principles of operation, laser types, and laser emissions
(2) Laser safety, including the following:
   a. System operating procedures
   b. Risk assessment and risk control procedures
   c. Need for personnel protection
   d. Accident reporting procedures
   e. Biological effects of the laser upon the eye and the skin
   f. Electrical and other hazards associated with the laser equipment, including the following:
      i. High voltages (>1 kV) and stored energy in the capacitor banks
      ii. Circuit components, such as electron tubes, with anode voltages greater than 5 kV emitting X-rays
      iii. Capacitor bank explosions
      iv. Production of ionizing radiation
      v. Poisoning from the solvent or dye switching liquids or laser media
      vi. High sound intensity levels from pulsed lasers

(C) Proof of Qualification. Proof of qualification of the laser equipment operator shall be readily available.

330.4 Safeguarding of Employees in the Laser Operating Area.

(A) Eye Protection. Employees shall be provided with eye protection as required by federal regulation.

(B) Warning Signs. Warning signs shall be posted at the entrances to areas or protective enclosures containing laser products.

(C) Master Control. High-power laser equipment shall include a key-operated master control.

(D) High-Power Radiation Emission Warning. High-power laser equipment shall include a fail-safe laser radiation emission audible and visible warning when it is switched on or if the capacitor banks are charged.

(E) Beam Shutters or Caps. Beam shutters or caps shall be used, or the laser switched off, when laser transmission is not required. The laser shall be switched off when unattended for 30 minutes or more.

(F) Aiming. Laser beams shall not be aimed at employees.

(G) Label. Laser equipment shall bear a label indicating its maximum output.

(H) Personal Protective Equipment (PPE). PPE shall be provided for users and operators of high-power laser equipment.

330.5 Employee Responsibility. Employees shall be responsible for the following:

(1) Obtaining authorization for laser use
(2) Obtaining authorization for being in a laser operating area
(3) Observing safety rules
(4) Reporting laser equipment failures and accidents to the employer

ARTICLE 340 Safety-Related Work Practices: Power Electronic Equipment

340.1 Scope. This article shall apply to safety-related work practices around power electronic equipment, including the following:

(1) Electric arc welding equipment
(2) High-power radio, radar, and television transmitting towers and antennas
(3) Industrial dielectric and radio frequency (RF) induction heaters
(4) Shortwave or RF diathermy devices
(5) Process equipment that includes rectifiers and inverters such as the following:
   a. Motor drives
   b. Uninterruptible power supply systems
   c. Lighting controllers

340.2 Definition. For the purposes of this article, the definition that follows shall apply.

Radiation Worker. A person who is required to work in electromagnetic fields, the radiation levels of which exceed those specified for nonoccupational exposure.

340.3 Application. The purpose of this article is to provide guidance for safety personnel in preparing specific safety-related work practices within their industry.

340.4 Reference Standards. The following are reference standards for use in the preparation of specific guidance to employees as follows:

(1) International Electrotechnical Commission IEC 60479, *Effects of Current Passing Through the Body*:
   a. 60479-1 Part 1: General aspects
   b. 60479-1-1 Chapter 1: Electrical impedance of the human body
   c. 60479-1-2 Chapter 2: Effects of ac in the range of 15 Hz to 100 Hz
   d. 60479-2 Part 2: Special aspects
   e. 60479-2-4 Chapter 4: Effects of ac with frequencies above 100 Hz
   f. 60479-2-5 Chapter 5: Effects of special waveforms of current
   g. 60479-2-6 Chapter 6: Effects of unidirectional single impulse currents of short duration
(2) International Commission on Radiological Protection (ICRP) Publication 33, *Protection Against Ionizing Radiation from External Sources Used in Medicine*

340.5 Effects of Electricity on the Human Body. The employer and employees shall be aware of the following hazards associated with power electronic equipment.

(1) Effects of Power Frequency Current:
   a. At 0.5 mA, shock is perceptible.
   b. At 10 mA, a person may not be able to voluntarily let go of an energized electrical conductor or circuit part.
   c. At about 40 mA, the shock, if lasting for 1 second or longer, can be fatal due to ventricular fibrillation.
   d. Further increasing current leads to burns and cardiac arrest.

(2) Effects of Direct Current:
   a. A dc current of 2 mA is perceptible.
   b. A dc current of 40 mA is considered the threshold of the let-go current.
(3) Effects of Voltage. A voltage of 30 V rms, or 60 V dc, is considered safe, except when the skin is broken. The internal body resistance can be as low as 500 ohms, so fatalities can occur.
(4) Effects of Short Contact:
   a. For contact less than 0.1 second and with currents just greater than 0.5 mA, ventricular fibrillation can occur only if the shock is during a vulnerable part of the cardiac cycle.
   b. For contact of less than 0.1 second and with currents of several amperes, ventricular fibrillation can occur if the shock is during a vulnerable part of the cardiac cycle.
   c. For contact of greater than 0.8 second and with currents just greater than 0.5 A, cardiac arrest (reversible) can occur.
   d. For contact greater than 0.8 second and with currents of several amperes, burns and death are probable.
(5) Effects of Alternating Current at Frequencies Above 100 Hz. When the threshold of perception increases from 10 kHz to 100 kHz, the threshold of let-go current increases from 10 mA to 100 mA.
(6) Effects of Waveshape. Contact with voltages from phase controls usually causes effects between those of ac and dc sources.
(7) Effects of Capacitive Discharge:
   a. A circuit of capacitance of 1 μF having a 10 kV capacitor charge can cause ventricular fibrillation.
   b. A circuit of capacitance of 20 μF having a 10 kV capacitor charge can be dangerous and probably will cause ventricular fibrillation.

340.6 Hazards Associated with Power Electronic Equipment. The employer and employees shall be aware of the hazards associated with the following:

(1) High voltages within the power supplies
(2) Radio frequency energy–induced high voltages
(3) Effects of RF fields in the vicinity of antennas and antenna transmission lines, which can introduce electrical shock and burns
(4) Ionizing (X-radiation) hazards from magnetrons, klystrons, thyratrons, cathode-ray tubes, and similar devices
(5) Nonionizing RF radiation hazards from the following:
   a. Radar equipment
   b. Radio communication equipment, including broadcast transmitters
   c. Satellite–earth-transmitters
   d. Industrial scientific and medical equipment
   e. RF induction heaters and dielectric heaters
   f. Industrial microwave heaters and diathermy radiators

340.7 Specific Measures for Personnel Safety.

(A) Employer Responsibility. The employer shall be responsible for the following:

(1) Proper training and supervision by properly qualified personnel, including the following:
   a. Identification of associated hazards
   b. Strategies to reduce the risk associated with the hazards
   c. Methods of avoiding or protecting against the hazard
   d. Necessity of reporting any incident that resulted in, or could have resulted in, injury or damage to health
(2) Properly installed equipment
(3) Proper access to the equipment
(4) Availability of the correct tools for operation and maintenance
(5) Proper identification and guarding of dangerous equipment
(6) Provision of complete and accurate circuit diagrams and other published information to the employee prior to the employee starting work (The circuit diagrams should be marked to indicate the components that present an electrical hazard.)
(7) Maintenance of clear and clean work areas around the equipment to be worked on
(8) Provision of adequate and proper illumination of the work area

(B) Employee Responsibility. The employee shall be responsible for the following:

(1) Understanding the hazards associated with the work
(2) Being continuously alert and aware of the possible hazards
(3) Using the proper tools and procedures for the work
(4) Informing the employer of malfunctioning protective measures, such as faulty or inoperable enclosures and locking schemes
(5) Examining all documents provided by the employer relevant to the work, especially those documents indicating the location of components that present an electrical hazard
(6) Maintaining good housekeeping around the equipment and work space
(7) Reporting any incident that resulted in, or could have resulted in, injury or damage to health
(8) Using and appropriately maintaining the PPE and tools required to perform the work safely

ARTICLE 350
Safety-Related Work Requirements: Research and Development Laboratories

350.1 Scope. The requirements of this article shall apply to the electrical installations in those areas, with custom or special electrical equipment, designated by the facility management for research and development (R&D) or as laboratories.

350.2 Definitions. For the purposes of this article, the definitions that follow shall apply.

Competent Person. A person who meets all the requirements of *qualified person*, as defined in Article 100 in Chapter 1 of this standard and who, in addition, is responsible for all work activities or safety procedures related to custom or special equipment and has detailed knowledge regarding the exposure to electrical hazards, the appropriate control methods to reduce the risk associated with those hazards, and the implementation of those methods.

Field Evaluated. A thorough evaluation of nonlisted or modified equipment in the field that is performed by persons or parties acceptable to the authority having jurisdiction. The evaluation approval ensures that the equipment meets appropriate codes and standards, or is similarly found suitable for a specified purpose.

Laboratory. A building, space, room, or group of rooms intended to serve activities involving procedures for investigation, diagnostics, product testing, or use of custom or special electrical components, systems, or equipment.

Research and Development (R&D). An activity in an installation specifically designated for research or development conducted with custom or special electrical equipment.

350.3 Applications of Other Articles. The electrical system for R&D and laboratory applications shall meet the requirements of the remainder of this document, except as amended by Article 350.

Informational Note: Examples of these applications include low-voltage–high-current power systems; high-voltage–low-current power systems; dc power supplies; capacitors; cable trays for signal cables and other systems, such as steam, water, air, gas, or drainage; and custom-made electronic equipment.

350.4 Specific Measures and Controls for Personnel Safety. Each laboratory or R&D system application shall be assigned a competent person as defined in this article to ensure the use of appropriate electrical safety-related work practices and controls.

350.5 Listing Requirements. The equipment or systems used in the R&D area or in the laboratory shall be listed or field evaluated prior to use.

Informational Note: Laboratory and R&D equipment or systems can pose unique electrical hazards that might require mitigation. Such hazards include ac and dc, low voltage and high amperage, high voltage and low current, large electromagnetic fields, induced voltages, pulsed power, multiple frequencies, and similar exposures.

Informative Annex A Referenced Publications

A.1 General. This informative annex is not part of the requirements of this document and is included for information only. To the extent the documents or portions thereof listed in this informative annex are referenced within this standard, those documents are to be considered part of the requirements of this document in the section and manner in which they are referenced.

A.2 NFPA Publications. National Fire Protection Association, 1 Batterymarch Park, Quincy, MA 02169-7471.

NFPA 70®, *National Electrical Code®*, 2014 edition.

A.3 Other Publications.

A.3.1 ANSI Publications. American National Standards Institute, Inc., 25 West 43rd Street, 4th Floor, New York, NY 10036.

ANSI/ASC A14.1, *American National Standard for Ladders — Wood — Safety Requirements,* 2007.

ANSI/ASC A14.3, *American National Standard for Ladders — Fixed — Safety Requirements,* 2008.

ANSI/ASC A14.4, *American National Standard Safety Requirements for Job-Made Ladders,* 2009.

ANSI/ASC A14.5, *American National Standard for Ladders — Portable Reinforced — Safety Requirements,* 2007.

ANSI Z87.1, *Practice for Occupational and Educational Eye and Face Protection,* 2010.

ANSI Z89.1, *Requirements for Protective Headwear for Industrial Workers,* 2009.

ANSI Z535, *Series of Standards for Safety Signs and Tags,* 2011.

A.3.2 ASTM Publications. ASTM International, 100 Barr Harbor Drive, P.O Box C700, West Conshohocken, PA 19428-2959.

ASTM D120, *Standard Specification for Rubber Insulating Gloves*, 2009.

ASTM D1048, *Standard Specification for Rubber Insulating Blankets,* 2012.

ASTM D1049, *Standard Specification for Rubber Covers,* 1998 (R 2010).

ASTM D1050, *Standard Specification for Rubber Insulating Line Hoses,* 2005 (R 2011).

ASTM D1051, *Standard Specification for Rubber Insulating Sleeves*, 2008.

ASTM F478, *Standard Specification for In-Service Care of Insulating Line Hose and Covers*, 2009.

ASTM F479, *Standard Specification for In-Service Care of Insulating Blankets,* 2006 (R 2011).

ASTM F496, *Standard Specification for In-Service Care of Insulating Gloves and Sleeves*, 2008.

ASTM F696, *Standard Specification for Leather Protectors for Rubber Insulating Gloves and Mittens,* 2006 (R 2011).

ASTM F711, *Standard Specification for Fiberglass-Reinforced Plastic (FRP) Rod and Tube Used in Live Line Tools,* 2002 (R 2007).

ASTM F712, *Standard Test Methods and Specifications for Electrically Insulating Plastic Guard Equipment for Protection of Workers,* 2006 (R 2011).

ASTM F855, *Standard Specification for Temporary Protective Grounds to Be Used on De-energized Electric Power Lines and Equipment*, 2009.

ASTM F887, *Standard Specification for Personal Climbing Equipment*, 2011.

ASTM F1116, *Standard Test Method for Determining Dielectric Strength of Dielectric Footwear*, 2003 (R 2008).

ASTM F1117, *Standard Specification for Dielectric Footwear*, 2003 (R 2008).

ASTM F1236, *Standard Guide for Visual Inspection of Electrical Protective Rubber Products,* 1996 (R 2012).

ASTM F1296, *Standard Guide for Evaluating Chemical Protective Clothing*, 2008.

ASTM F1449, *Standard Guide for Industrial Laundering of Flame, Thermal, and Arc Resistant Clothing*, 2008.

ASTM F1505, *Standard Specification for Insulated and Insulating Hand Tools,* 2010.

ASTM F1506, *Standard Performance Specification for Flame Resistant and Arc Rated Textile Materials for Wearing Apparel for Use by Electrical Workers Exposed to Momentary Electric Arc and Related Thermal Hazards*, 2010a.

ASTM F1742, *Standard Specification for PVC Insulating Sheeting*, 2003 (R 2011).

ASTM F1891, *Standard Specification for Arc and Flame Resistant Rainwear,* 2012.

ASTM F1959/F1959M, *Standard Test Method for Determining the Arc Rating of Materials for Clothing,* 2012.

ASTM F2178, *Standard Test Method for Determining the Arc Rating and Standard Specification for Eye or Face Protective Products*, 2012.

ASTM F2249, *Standard Specification for In-Service Test Methods for Temporary Grounding Jumper Assemblies Used on De-Energized Electric Power Lines and Equipment*, 2003 (R 2009).

ASTM F2412/F2320, *Standard Specification for Rubber Insulating Sheeting*, 2011.

ASTM F2412, *Standard Test Methods for Foot Protections*, 2011.

ASTM F2413, *Standard Specification for Performance Requirements for Protective (Safety) Toe Cap Footwear*, 2011.

ASTM F2522, *Standard Test Method for Determining the Protective Performance of a Shield Attached on Live Line Tools or on Racking Rods for Electric Arc Hazards*, 2012.

ASTM F2676, *Standard Test Method for Determining the Protective Performance of an Arc Protective Blanket for Electric Arc Hazards*, 2009.

ASTM F2677, *Standard Specification for Electrically Insulating Aprons*, 2008a.

ASTM F2757, *Standard Guide for Home Laundering Care and Maintenance of Flame, Thermal and Arc Resistant Clothing*, 2009.

A.3.3 ICRP Publications. International Commission on Radiological Protection, SE-171 16 Stockholm, Sweden.

ICRP Publication 33, *Protection Against Ionizing Radiation from External Sources Used in Medicine*, March 1981.

A.3.4 IEC Publications. International Electrotechnical Commission, 3, rue de Varembé, P.O. Box 131, CH-1211 Geneva 20, Switzerland.

IEC 60479, *Effects of Current Passing Through the Body*.

60479-1 Part 1: General aspects, 2005.

60479-1-1 Chapter 1: Electrical impedance of the human body.

60479-1-2 Chapter 2: Effects of ac in the range of 15 Hz to 100 Hz.

60479-2 Part 2: Special aspects, 2007.

60479-2-4 Chapter 4: Effects of ac with frequencies above 100 Hz.

60479-2-5 Chapter 5: Effects of special waveforms of current.

60479-2-6 Chapter 6: Effects of unidirectional single impulse currents of short duration.

A.3.5 IEEE Publications. Institute of Electrical and Electronics Engineers, IEEE Operations Center, 445 Hoes Lane, P.O. Box 1331, Piscataway, NJ 08855-1331.

IEEE C37.20.7, *Guide for Testing Metal-Enclosed Switchgear Rated up to 38 kV for Internal Arcing Faults*, 2007/Corrigendum 1, 2010.

A.4 References for Extracts in Mandatory Sections.

NFPA 70®, *National Electrical Code®*, 2014 edition.

Informative Annex B Informational References

B.1 Referenced Publications. The following documents or portions thereof are referenced within this standard for informational purposes only and are thus not part of the requirements of this document unless also listed in Informative Annex A.

B.1.1 NFPA Publications. National Fire Protection Association, 1 Batterymarch Park, Quincy, MA 02169-7471.

NFPA 1, *Fire Code*, 2015 edition.

NFPA 70,® National Electrical Code®, 2014 edition.

NFPA 70B, *Recommended Practice for Electrical Equipment Maintenance,* 2013 edition.

NFPA 79, *Electrical Standard for Industrial Machinery*, 2015 edition.

B.1.2 ANSI Publications. American National Standards Institute, Inc., 25 West 43rd Street, 4th Floor, New York, NY 10036.

ANSI/AIHA Z10, *American National Standard for Occupational Health and Safety Management Systems*, 2012.

ANSI/ASSE Z244.1, *Control of Hazardous Energy — Lockout/Tagout and Alternative Methods*, 2003 (R 2008).

ANSI C84.1, *Electric Power Systems and Equipment – Voltage Ratings (60 Hz)*, 2011.

ANSI/ISO 14001, *Environmental Management Systems — Requirements with Guidance for Use*, 2004/Corrigendum 1, 2009.

ANSI/NETA MTS, *Standard for Maintenance Testing Specifications for Electrical Power Distribution Equipment and Systems,* 2011.

ANSI Z535.4, *Product Safety Signs and Labels*, 2011.

B.1.3 ASTM Publications. ASTM International, 100 Barr Harbor Drive, P.O. Box C700, West Conshohocken, PA 19428-2959.

ASTM F496, *Standard Specification for In-Service Care of Insulating Gloves and Sleeves*, 2008.

ASTM F711, *Standard Specification for Fiberglass-Reinforced Plastic (FRP) Rod and Tube Used in Live Line Tools*, 2002 (R 2007).

ASTM F1449, *Standard Guide for Industrial Laundering of Flame, Thermal, and Arc Resistant Clothing*, 2008.

ASTM F1506, *Standard Performance Specification for Flame Resistant and Arc Rated Textile Materials for Wearing Apparel for Use by Electrical Workers Exposed to Momentary Electric Arc and Related Thermal Hazards,* 2010a.

ASTM F1959/F1959M, *Standard Test Method for Determining the Arc Rating of Materials for Clothing*, 2012.

ASTM F2249, *Standard Specification for In-Service Test Methods for Temporary Grounding Jumper Assemblies Used on De-Energized Electric Power Lines and Equipment*, 2003 (R 2009).

ASTM F2413, *Standard Specifications for Performance Requirements for Protective (Safety) Toe Cap Footwear,* 2011.

ASTM F2757, *Standard Guide for Home Laundering Care and Maintenance of Flame, Thermal and Arc Resistant Clothing*, 2009.

B.1.4 British Standards Institute, Occupational Health and Safety Assessment Series (OHSAS) Project Group Publications. British Standards Institute, American Headquarters, 12110 Sunset Hills Road, Suite 200, Reston VA 20190-5902.

BS OSHAS 18001, *Occupational Health and Safety Management Systems*, 2007.

B.1.5 CSA Publications. Canadian Standards Association, 5060 Spectrum Way, Mississauga, ON L4W 5N6, Canada.

CAN/CSA Z462, *Workplace Electrical Safety*, 2012.

CAN/CSA Z1000, *Occupational Health and Safety Management*, 2006 (R 2011).

B.1.6 IEC Publications. International Electrotechnical Commission, 3, rue de Varembé, P.O. Box 131, CH-1211 Geneva 20, Switzerland.

IEC 60204-1 ed 5.1 Consol. with am 1, *Safety of Machinery — Electrical Equipment of Machines — Part 1: General Requirements,* 2009.

B.1.7 IEEE Publications. Institute of Electrical and Electronic Engineers, IEEE Operations Center, 445 Hoes Lane, P.O. Box 1331, Piscataway, NJ 08855-1331.

ANSI/IEEE C2, *National Electrical Safety Code,* 2012.

ANSI/IEEE C 37.20.6, *Standard for 4.76 kV to 38 kV-Rated Ground and Test Devices Used in Enclosures*, 2007.

IEEE 4, *Standard Techniques for High Voltage Testing*, 2013.

IEEE 450, *IEEE Recommended Practice for Maintenance, Testing, and Replacement of Vented Lead-Acid Batteries for Stationary Applications*, 2010.

IEEE 516, *Guide for Maintenance Methods on Energized Power Lines*, 2009.

IEEE 937, *Recommended Practice for Installation and Maintenance of Lead-Acid Batteries for Photovoltaic Systems,* 2007.

IEEE 946, *IEEE Recommended Practice for the Design of DC Auxiliary Power Systems for Generating Systems,* 2004.

IEEE 1106, *IEEE Recommended Practice for Installation, Maintenance, Testing, and Replacement of Vented Nickel-Cadmium Batteries for Stationary Applications*, 2005 (R 2011).

IEEE 1184, *IEEE Guide for Batteries for Uninterruptible Power Supply Systems*, 2006.

IEEE 1187, *Recommended Practice for Installation Design and Installation of Valve-Regulated Lead-Acid Storage Batteries for Stationary Applications,* 2002.

IEEE 1188, *IEEE Recommended Practice for Maintenance, Testing, and Replacement of Valve-Regulated Lead-Acid (VRLA) Batteries for Stationary Applications*, 2005 (R 2010).

• IEEE 1491, *IEEE Guide for Selection and Use of Battery Monitoring Equipment in Stationary Applications*, 2012.

IEEE 1584™, *Guide for Performing Arc Flash Hazard Calculations,* 2002.

IEEE 1584a™, *Guide for Performing Arc Flash Hazard Calculations, Amendment 1*, 2004.

IEEE 1584b™, *Guide for Performing Arc Flash Hazard Calculations — Amendment 2: Changes to Clause 4*, 2011.

IEEE 1657, *Recommended Practice for Personnel Qualifications for Installation and Maintenance of Stationary Batteries,* 2009.

IEEE 3007.1, *IEEE Recommended Practice for the Operation and Management of Industrial and Commercial Power Systems*, 2010.

IEEE 3007.2, *IEEE Recommended Practice for the Maintenance of Industrial and Commercial Power Systems*, 2010.

IEEE 3007.3, *IEEE Recommended Practice for Electrical Safety in Industrial and Commercial Power Systems*, 2012.

Anderson, W. E., "Risk Analysis Methodology Applied to Industrial Machine Development," *IEEE Transactions on Industrial Applications,* Vol. 41, No. 1, January/February 2005, pp. 180–187.

Ammerman, R. F., Gammon, T., Sen, P. K., and Nelson, J. P., "DC-Arc Models and Incident-Energy Calculations," *IEEE Transactions on Industrial Applications*, Vol. 46, No. 5, 2010.

Doan, D. R, "Arc Flash Calculations for Exposures to DC Systems," *IEEE Transactions on Industrial Applications*, Vol 46, No. 6, 2010.

Doughty, R. L., T. E. Neal, and H. L. Floyd II, "Predicting Incident Energy to Better Manage the Electric Arc Hazard on 600 V Power Distribution Systems," *Record of Conference Papers IEEE IAS 45th Annual Petroleum and Chemical Industry Conference*, September 28–30, 1998.

Lee, R., "The Other Electrical Hazard: Electrical Arc Flash Burns," *IEEE Trans. Applications*, Vol. 1A-18, No. 3, May/June 1982.

B.1.8 ISA Publications. Instrumentation, Systems, and Automation Society, 67 Alexander Drive, Research Triangle Park, NC 27709.

ANSI/ISA 61010-1, *Safety Requirements for Electrical Equipment for Measurement, Control, and Laboratory Use, Part 1: General Requirements,* 2007.

B.1.9 ISEA Publications. International Safety Equipment Association, 1901 North Moore Street, Arlington, VA 22209-1762.

ANSI/ISEA Z358.1, *American National Standard for Emergency Eye Wash and Shower Equipment*, 2009.

B.1.10 ISO Publications. International Organization for Standardization, 1, Ch. de la Voie-Creuse, Case postale 56, CH-1211 Geneva 20, Switzerland.

ISO 14001, *Environmental Management Systems — Requirements with Guidance for Use*, 2004.

B.1.11 NIOSH Publications. National Institute for Occupational Safety and Health, Centers for Disease Control and Prevention, 1600 Clifton Road, Atlanta, GA 30333.

DHHS (NIOSH) Publication No. 94-110, *Applications Manual for the Revised NIOSH Lifting Equation*, 1994.

B.1.12 UL Publications. Underwriters Laboratories Inc., 333 Pfingsten Road, Northbrook, IL 60062-2096.

ANSI/UL 943, *Standard for Ground-Fault Circuit Interrupters*, 2006 (R 2012).

•

B.1.13 U.S. Government Publications. U.S. Government Printing Office, Washington, DC 20402.

• Title 29, Code of Federal Regulations, Part 1910, Occupational Safety and Health Standards, Subpart S, Electrical, 1910.137, Personal Protective Equipment, and 1910.305(j)(7), Storage Batteries; and Part 1926, Safety and Health Regulations for Construction, Subpart K, Electrical, 1926.441, Batteries and Battery Charging.

B.1.14 Other Publications.

• "DC Arc Hazard Assessment Phase II," Copyright Material, Kinectrics Inc., Report No. K-012623-RA-0002-R00.

Informative Annex C Limits of Approach

This informative annex is not a part of the requirements of this NFPA document but is included for informational purposes only.

C.1 Preparation for Approach. Observing a safe approach distance from exposed energized electrical conductors or circuit parts is an effective means of maintaining electrical safety. As the distance between a person and the exposed energized conductors or circuit parts decreases, the potential for electrical accident increases.

C.1.1 Unqualified Persons, Safe Approach Distance. Unqualified persons are safe when they maintain a distance from the exposed energized conductors or circuit parts, including the longest conductive object being handled, so that they cannot contact or enter a specified air insulation distance to the exposed energized electrical conductors or circuit parts. This safe approach distance is the limited approach boundary. Further, persons must not cross the arc flash boundary unless they are wearing appropriate personal protective clothing and are under the close supervision of a qualified person. Only when continuously escorted by a qualified person should an unqualified person cross the limited approach boundary. Under no circumstance should an unqualified person cross the restricted approach boundary, where special shock protection techniques and equipment are required.

C.1.2 Qualified Persons, Safe Approach Distance.

C.1.2.1 Determine the arc flash boundary and, if the boundary is to be crossed, appropriate arc-rated protective equipment must be utilized.

C.1.2.2 For a person to cross the limited approach boundary and enter the limited space, a person should meet the following criteria:

(1) Be qualified to perform the job/task
(2) Be able to identify the hazards and associated risks with the tasks to be performed

C.1.2.3 To cross the restricted approach boundary and enter the restricted space, qualified persons should meet the following criteria:

(1) Have an energized electrical work permit authorized by management
(2) Use personal protective equipment (PPE) that is rated for the voltage and energy level involved
(3) Minimize the likelihood of bodily contact with exposed energized conductors and circuit parts from inadvertent movement by keeping as much of the body out of the restricted space as possible and using only protected body parts in the space as necessary to accomplish the work
(4) Use insulated tools and equipment

(See Figure C.1.2.3.)

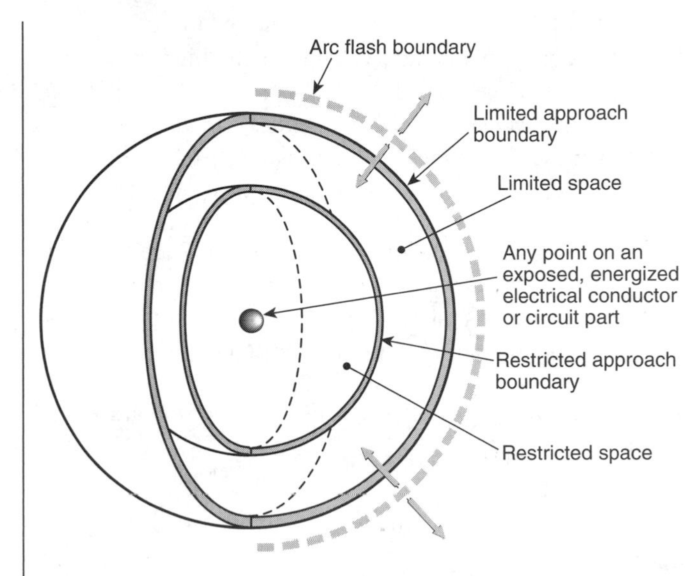

Figure C.1.2.3 Limits of Approach.

C.2 Basis for Distance Values in Tables 130.4(D)(a) and 130.4(D)(b).

C.2.1 General Statement. Columns 2 through 4 of Table 130.4(D)(a) and Table 130.4(D)(b) show various distances from the exposed energized electrical conductors or circuit parts. They include dimensions that are added to a basic minimum air insulation distance. Those basic minimum air insulation distances for voltages 72.5 kV and under are based on IEEE 4, *Standard Techniques for High Voltage Testing,* Appendix 2B; and voltages over 72.5 kV are based on IEEE 516, *Guide for Maintenance Methods on Energized Power Lines*. The minimum air insulation distances that are required to avoid flashover are as follows:

(1) ≤300 V: 1 mm (0 ft 0.03 in.)
(2) >300 V to ≤750 V: 2 mm (0 ft 0.07 in.)
(3) >750 V to ≤2 kV: 5 mm (0 ft 0.19 in.)
(4) >2 kV to ≤15 kV: 39 mm (0 ft 1.5 in.)
(5) >15 kV to ≤36 kV: 161 mm (0 ft 6.3 in.)
(6) >36 kV to ≤48.3 kV: 254 mm (0 ft 10.0 in.)
(7) >48.3 kV to ≤72.5 kV: 381 mm (1 ft 3.0 in.)
(8) >72.5 kV to ≤121 kV: 640 mm (2 ft 1.2 in.)
(9) >138 kV to ≤145 kV: 778 mm (2 ft 6.6 in.)

(10) >161 kV to ≤169 kV: 915 mm (3 ft 0.0 in.)

(11) >230 kV to ≤242 kV: 1.281 m (4 ft 2.4 in.)

(12) >345 kV to ≤362 kV: 2.282 m (7 ft 5.8 in.)

(13) >500 kV to ≤550 kV: 3.112 m (10 ft 2.5 in.)

(14) >765 kV to ≤800 kV: 4.225 m (13 ft 10.3 in.)

C.2.1.1 Column 1. The voltage ranges have been selected to group voltages that require similar approach distances based on the sum of the electrical withstand distance and an inadvertent movement factor. The value of the upper limit for a range is the maximum voltage for the highest nominal voltage in the range, based on ANSI C84.1, *Electric Power Systems and Equipment— Voltage Ratings (60 Hz).* For single-phase systems, select the range that is equal to the system's maximum phase-to-ground voltage multiplied by 1.732.

C.2.1.2 Column 2. The distances in column 2 are based on OSHA's rule for unqualified persons to maintain a 3.05 m (10 ft) clearance for all voltages up to 50 kV (voltage-to-ground), plus 100 mm (4.0 in.) for each 10 kV over 50 kV.

C.2.1.3 Column 3. The distances in column 3 are based on the following:

(1) ≤750 V: Use *NEC* Table 110.26(A)(1), Working Spaces, Condition 2, for the 151 V to 600 V range.

(2) >750 V to ≤145 kV: Use *NEC* Table 110.34(A), Working Space, Condition 2.

(3) >145 kV: Use OSHA's 3.05 m (10 ft) rules as used in Column 2.

C.2.1.4 Column 4. The distances in column 4 are based on adding to the flashover dimensions shown in C.2.1 the following inadvertent movement distance:

≤300 V: Avoid contact.

Based on experience and precautions for household 120/240-V systems:

>300 V to ≤750 V: Add 304.8 mm (1 ft 0 in.) for inadvertent movement.

These values have been found to be adequate over years of use in ANSI/IEEE C2, *National Electrical Safety Code,* in the approach distances for communication workers.

>72.5 kV: Add 304.8 mm (1 ft 0 in.) for inadvertent movement.

These values have been found to be adequate over years of use in ANSI/IEEE C2, *National Electrical Safety Code,* in the approach distances for supply workers.

Informative Annex D Incident Energy and Arc Flash Boundary Calculation Methods

This informative annex is not a part of the requirements of this NFPA document but is included for informational purposes only.

D.1 Introduction. Informative Annex D summarizes calculation methods available for calculating arc flash boundary and incident energy. It is important to investigate the limitations of any methods to be used. The limitations of methods summarized in Informative Annex D are described in Table D.1.

D.2 Ralph Lee Calculation Method.

D.2.1 Basic Equations for Calculating Arc Flash Boundary Distances. The short-circuit symmetrical ampacity, I_{sc}, from a bolted three-phase fault at the transformer terminals is calculated with the following formula:

[D.2.1(a)]

$$I_{sc} = \{[MVA \text{ Base } \times 10^6] \div [1.732 \times V]\} \times \{100 \div \%Z\}$$

where I_{sc} is in amperes, V is in volts, and $\%Z$ is based on the transformer *MVA*.

A typical value for the maximum power, P (in MW) in a three-phase arc can be calculated using the following formula:

[D.2.1(b)]

$$P = [\text{maximum bolted fault, in } MVA_{bf}] \times 0.707^2$$

[D.2.1(c)]

$$P = 1.732 \times V \times I_{sc} \times 10^{-6} \times 0.707^2$$

The arc flash boundary distance is calculated in accordance with the following formulae:

[D.2.1(d)]

$$D_c = [2.65 \times MVA_{bf} \times t]^{1/2}$$

[D.2.1(e)]

$$D_c = [53 \times MVA \times t]^{1/2}$$

where:

D_c = distance in feet of person from arc source for a just curable burn (that is, skin temperature remains less than 80°C).

MVA_{bf} = bolted fault *MVA* at point involved.

MVA = *MVA* rating of transformer. For transformers with *MVA* ratings below 0.75 *MVA*, multiply the transformer *MVA* rating by 1.25.

t = time of arc exposure in seconds.

The clearing time for a current-limiting fuse is approximately ¼ cycle or 0.004 second if the arcing fault current is in the fuse's current-limiting range. The clearing time of a 5-kV and 15-kV circuit breaker is approximately 0.1 second or 6 cycles if the instantaneous function is installed and operating. This can be broken down as follows: actual breaker time (approximately 2 cycles), plus relay operating time of approximately 1.74 cycles, plus an additional safety margin of 2 cycles, giving a total time of approximately 6 cycles. Additional time must be added if a time delay function is installed and operating.

The formulas used in this explanation are from Ralph Lee, "The Other Electrical Hazard: Electrical Arc Flash Burns," in *IEEE Trans. Industrial Applications*. The calculations are based on the worst-case arc impedance. *(See Table D.2.1.)*

Table D.1 Limitation of Calculation Methods

Section	Source	Limitations/Parameters
D.2	Lee, "The Other Electrical Hazard: Electrical Arc Flash Burns"	Calculates incident energy and arc flash boundary for arc in open air; conservative over 600 V and becomes more conservative as voltage increases
D.3	Doughty, et al., "Predicting Incident Energy to Better Manage the Electrical Arc Hazard on 600 V Power Distribution Systems"	Calculates incident energy for three-phase arc on systems rated 600 V and below; applies to short-circuit currents between 16 kA and 50 kA
D.4	IEEE 1584, *Guide for Performing Arc Flash Calculations*	Calculates incident energy and arc flash boundary for: 208 V to 15 kV; three-phase; 50 Hz to 60 Hz; 700 A to 106,000 A short-circuit current; and 13 mm to 152 mm conductor gaps
D.5	Doan, "Arc Flash Calculations for Exposure to DC Systems"	Calculates incident energy for dc systems rated up to 1000 V dc

Table D.2.1 Flash Burn Hazard at Various Levels in a Large Petrochemical Plant

(1)	(2)	(3)	(4)	(5)	(6)	(7)	
Bus Nominal Voltage Levels	System (MVA)	Transformer (MVA)	System or Transformer (% Z)	Short-Circuit Symmetrical (A)	Clearing Time of Fault (cycles)	Arc Flash Boundary Typical Distance*	
						SI	U.S.
230 kV	9000		1.11	23,000	6.0	15 m	49.2 ft
13.8 kV	750		9.4	31,300	6.0	1.16 m	3.8 ft
Load side of all 13.8-V fuses	750		9.4	31,300	1.0	184 mm	0.61 ft
4.16 kV		10.0	5.5	25,000	6.0	2.96 m	9.7 ft
4.16 kV		5.0	5.5	12,600	6.0	1.4 m	4.6 ft
Line side of incoming 600-V fuse		2.5	5.5	44,000	60.0–120.0	7 m–11 m	23 ft–36 ft
600-V bus		2.5	5.5	44,000	0.25	268 mm	0.9 ft
600-V bus		1.5	5.5	26,000	6.0	1.6 m	5.4 ft
600-V bus		1.0	5.57	17,000	6.0	1.2 m	4 ft

*Distance from an open arc to limit skin damage to a curable second degree skin burn [less than 80°C (176°F) on skin] in free air.

D.2.2 Single-Line Diagram of a Typical Petrochemical Complex. The single-line diagram *(see Figure D.2.2)* illustrates the complexity of a distribution system in a typical petrochemical plant.

D.2.3 Sample Calculation. Many of the electrical characteristics of the systems and equipment are provided in Table D.2.1. The sample calculation is made on the 4160-volt bus 4A or 4B. Table D.2.1 tabulates the results of calculating the arc flash boundary for each part of the system. For this calculation, based on Table D.2.1, the following results are obtained:

(1) Calculation is made on a 4160-volt bus.
(2) Transformer *MVA* (and base *MVA*) = 10 *MVA*.
(3) Transformer impedance on 10 *MVA* base = 5.5 percent.
(4) Circuit breaker clearing time = 6 cycles.

Using Equation D.2.1(a), calculate the short-circuit current:

$$I_{sc} = \{[MVA \text{ Base } \times 10^6] \div [1.732 \times V]\} \times \{100 \div \%Z\}$$
$$= \{[10 \times 10^6] \div [1.732 \times 4160]\} \times \{100 \div 5.5\}$$
$$= 25{,}000 \text{ amperes}$$

Using Equation D.2.1(b), calculate the power in the arc:

$$P = 1.732 \times 4160 \times 25{,}000 \times 10^{-6} \times 0.707^2$$
$$= 91 \text{ MW}$$

Using Equation D.2.1(d), calculate the second degree burn distance:

$$D_c = \{2.65 \times [1.732 \times 25{,}000 \times 4160 \times 10^{-6}] \times 0.1\}^{1/2}$$
$$= 6.9 \text{ or } 7.00 \text{ ft}$$

Or, using Equation D.2.1(e), calculate the second degree burn distance using an alternative method:

$$D_c = [53 \times 10 \times 0.1]^{1/2}$$
$$= 7.28 \text{ ft}$$

D.2.4 Calculation of Incident Energy Exposure Greater Than 600 V for an Arc Flash Hazard Analysis. The equation that follows can be used to predict the incident energy produced by a three-phase arc in open air on systems rated above 600 V. The parameters required to make the calculations follow.

(1) The maximum bolted fault, three-phase short-circuit current available at the equipment.
(2) The total protective device clearing time (upstream of the prospective arc location) at the maximum short-circuit current. If the total protective device clearing time is longer than 2 seconds, consider how long a person is likely to remain in the location of the arc flash. It is likely that a person exposed to an arc flash will move away quickly if it is physically possible, and 2 seconds is a reasonable maximum time for calculations. A person in a bucket truck or a person who has

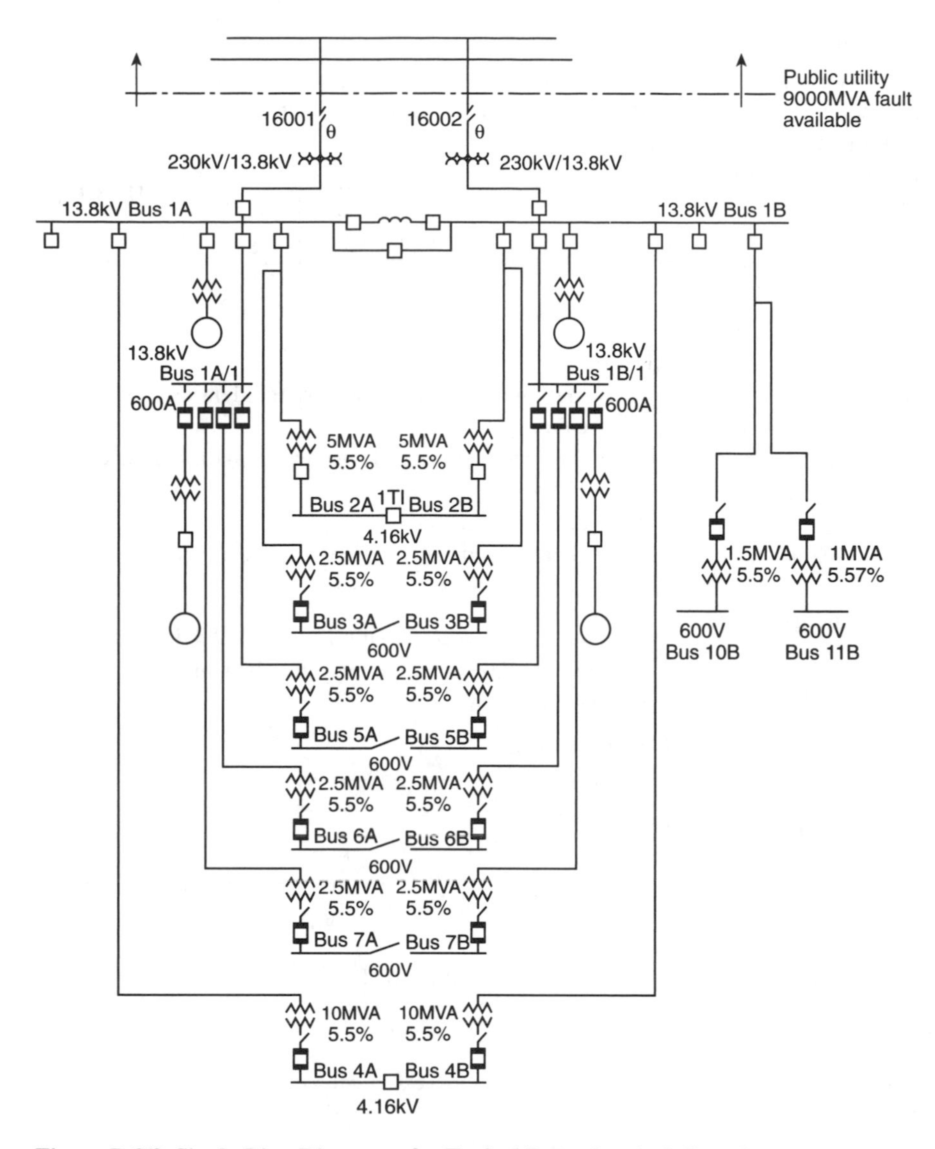

Figure D.2.2 Single-Line Diagram of a Typical Petrochemical Complex.

crawled into equipment will need more time to move away. Sound engineering judgment must be used in applying the 2-second maximum clearing time, since there could be circumstances where an employee's egress is inhibited.

(3) The distance from the arc source.

(4) Rated phase-to-phase voltage of the system.

$$E = \frac{793 \times F \times V \times t_A}{D^2} \quad \text{[D.2.4(4)]}$$

where:

E = incident energy, cal/cm^2
F = bolted fault short-circuit current, kA
V = system phase-to-phase voltage, kV
t_A = arc duration, sec
D = distance from the arc source, in.

D.3 Doughty Neal Paper.

D.3.1 Calculation of Incident Energy Exposure. The following equations can be used to predict the incident energy produced by a three-phase arc on systems rated 600 V and below. The results of these equations might not represent the worst case in all situations. It is essential that the equations be used only within the limitations indicated in the definitions of the variables shown under the equations. The equations must be used only under qualified engineering supervision.

Informational Note: Experimental testing continues to be performed to validate existing incident energy calculations and to determine new formulas.

The parameters required to make the calculations follow.

(1) The maximum bolted fault, three-phase short-circuit current available at the equipment and the minimum fault level at which the arc will self-sustain. (Calculations should be made using the maximum value, and then at lowest fault level at which the arc is self-sustaining. For 480-volt systems, the industry accepted minimum level for a sustaining arcing fault is 38 percent of the available bolted fault, three-phase short-circuit current. The highest incident energy exposure could occur at these lower levels where the overcurrent device could take seconds or minutes to open.)
(2) The total protective device clearing time (upstream of the prospective arc location) at the maximum short-circuit current, and at the minimum fault level at which the arc will sustain itself.
(3) The distance of the worker from the prospective arc for the task to be performed.

Typical working distances used for incident energy calculations are as follows:

(1) Low voltage (600 V and below) MCC and panelboards — 455 mm (18 in.)
(2) Low voltage (600 V and below) switchgear — 610 mm (24 in.)
(3) Medium voltage (above 600 V) switchgear — 910 mm (36 in.)

D.3.2 Arc in Open Air. The estimated incident energy for an arc in open air is as follows:

$$E_{MA} = 5271 D_A^{-1.9593} t_A \begin{bmatrix} 0.0016F^2 \\ -0.0076F \\ +0.8938 \end{bmatrix} \qquad \text{[D.3.2(a)]}$$

where:

E_{MA} = maximum open arc incident energy, cal/cm^2
D_A = distance from arc electrodes, in. (for distances 18 in. and greater)
t_A = arc duration, sec
F = short-circuit current, kA (for the range of 16 kA to 50 kA)

Sample Calculation: Using Equation D.3.2(a), calculate the maximum open arc incident energy, cal/cm^2, where D_A = 18 in., t_A = 0.2 second, and F = 20 kA.

$$E_{MA} = 5271\ D_A^{-1.9593} t_A \begin{bmatrix} 0.0016F^2 - 0.0076F \\ +0.8938 \end{bmatrix} \qquad \text{[D.3.2(b)]}$$
$$= 5271 \times .0035 \times 0.2[0.0016 \times 400 - 0.0076 \times 20 + 0.8938]$$
$$= 3.69 \times [1.381]$$
$$= 21.33\ \text{J/cm}^2 \left(5.098\ \text{cal/cm}^2\right)$$

D.3.3 Arc in a Cubic Box. The estimated incident energy for an arc in a cubic box (20 in. on each side, open on one end) is given in the equation that follows. This equation is applicable to arc flashes emanating from within switchgear, motor control centers, or other electrical equipment enclosures.

$$E_{MB} = 1038.7 D_B^{-1.4738} t_A \begin{bmatrix} 0.0093F^2 \\ -0.3453F \\ +5.9675 \end{bmatrix} \qquad \text{[D.3.3(a)]}$$

where:

E_{MB} = maximum 20 in. cubic box incident energy, cal/cm^2
D_B = distance from arc electrodes, in. (for distances 18 in. and greater)
t_A = arc duration, sec
F = short-circuit current, kA (for the range of 16 kA to 50 kA)

Sample Calculation: Using Equation D.3.3(a), calculate the maximum 20 in. cubic box incident energy, cal/cm^2, using the following:

(1) D_B = 18 in.
(2) t_A = 0.2 sec
(3) F = 20 kA

$$E_{MB} = 1038.7 D_B^{-1.4738} t_A \begin{bmatrix} 0.0093F^2 - 0.3453F \\ +5.9675 \end{bmatrix} \qquad \text{[D.3.3(b)]}$$
$$= 1038 \times 0.0141 \times 0.2 \begin{bmatrix} 0.0093 \times 400 - 0.3453 \times 20 \\ +5.9675 \end{bmatrix}$$
$$= 2.928 \times [2.7815]$$
$$= 34.1\ \text{J/cm}^2 \left(8.144\ \text{cal/cm}^2\right)$$

D.3.4 Reference. The equations for this section were derived in the IEEE paper by R. L. Doughty, T. E. Neal, and H. L. Floyd, II, "Predicting Incident Energy to Better Manage the Electric Arc Hazard on 600 V Power Distribution Systems."

D.4 IEEE 1584 Calculation Method.

D.4.1 Basic Equations for Calculating Incident Energy and Arc Flash Boundary. This section provides excerpts from IEEE 1584, *IEEE Guide for Performing Arc Flash Hazard Calculations,* for estimating incident energy and arc flash boundaries based on statistical analysis and curve fitting of available test data. An IEEE working group produced the data from tests it performed to produce models of incident energy.

The complete data, including a spreadsheet calculator to solve the equations, can be found in the IEEE 1584, *Guide for Performing Arc Flash Hazard Calculations.* Users are encouraged to consult the latest version of the complete document to understand the basis, limitation, rationale, and other pertinent information for proper application of the standard. It can be ordered from the Institute of Electrical and Electronics Engineers, Inc., 445 Hoes Lane, P.O. Box 1331, Piscataway, NJ 08855-1331.

D.4.1.1 System Limits. An equation for calculating incident energy can be empirically derived using statistical analysis of raw data along with a curve-fitting algorithm. It can be used for systems with the following limits:

(1) 0.208 kV to 15 kV, three-phase

(2) 50 Hz to 60 Hz

(3) 700 A to 106,000 A available short-circuit current

(4) 13 mm to 152 mm conductor gaps

For three-phase systems in open-air substations, open-air transmission systems, and distribution systems, a theoretically derived model is available. This theoretically derived model is intended for use with applications where faults escalate to three-phase faults. Where such an escalation is not possible or likely, or where single-phase systems are encountered, this equation will likely provide conservative results.

D.4.2 Arcing Current. To determine the operating time for protective devices, find the predicted three-phase arcing current.

For applications with a system voltage under 1 kV, solve Equation D.4.2(a) as follows:

$$\lg I_a = K + 0.662 \lg I_{bf} + 0.0966V + 0.000526G + 0.5588V\left(\lg I_{bf}\right) - 0.00304G\left(\lg I_{bf}\right) \quad \textbf{[D.4.2(a)]}$$

where:

lg = the $\log_{10}$

I_a = arcing current, kA

K = –0.153 for open air arcs; –0.097 for arcs-in-a-box

I_{bf} = bolted three-phase available short-circuit current (symmetrical rms), kA

V = system voltage, kV

G = conductor gap, mm *(see Table D.4.2)*

For systems greater than or equal to 1 kV, use Equation D.4.2(b):

$$\lg I_a = 0.00402 + 0.983 \lg I_{bf} \quad \textbf{[D.4.2(b)]}$$

This higher voltage formula is used for both open-air arcs and for arcs-in-a-box.

Convert from lg:

$$I_a = 10^{\lg I_a} \quad \textbf{[D.4.2(c)]}$$

Use 0.85 I_a to find a second arc duration. This second arc duration accounts for variations in the arcing current and the time for the overcurrent device to open. Calculate the incident energy using both arc durations (I_a and 0.85 I_a), and use the higher incident energy.

Table D.4.2 Factors for Equipment and Voltage Classes

System Voltage (kV)	Type of Equipment	Typical Conductor Gap (mm)	Distance Exponent Factor *x*
0.208–1	Open air	10–40	2.000
	Switchgear	32	1.473
	MCCs and panels	25	1.641
	Cables	13	2.000
>1–5	Open air	102	2.000
	Switchgear	13–102	0.973
	Cables	13	2.000
>5–15	Open air	13–153	2.000
	Switchgear	153	0.973
	Cables	13	2.000

D.4.3 Incident Energy at Working Distance — Empirically Derived Equation. To determine the incident energy using the empirically derived equation, determine the $\log_{10}$ of the normalized incident energy. The following equation is based on data normalized for an arc time of 0.2 second and a distance from the possible arc point to the person of 610 mm:

$$\lg E_n = k_1 + k_2 + 1.081 \lg I_a + 0.0011G \quad \textbf{[D.4.3(a)]}$$

where:

E_n = incident energy, normalized for time and distance, J/cm^2

k_1 = –0.792 for open air arcs

= –0.555 for arcs-in-a-box

k_2 = 0 for ungrounded and high-resistance grounded systems

= –0.113 for grounded systems

G = conductor gap, mm *(see Table D.4.2)*

Then,

$$E_n = 10^{\lg E_n} \quad \textbf{[D.4.3(b)]}$$

Converting from normalized:

$$E = 4.184 C_f \, E_n \left(\frac{t}{0.2} \right) \left(\frac{610^x}{D^x} \right) \quad \text{[D.4.3(c)]}$$

where:
E = incident energy, J/cm^2.
C_f = calculation factor
= 1.0 for voltages above 1 kV.
= 1.5 for voltages at or below 1 kV.
E_n = incident energy normalized.
t = arcing time, sec.
x = distance exponent from Table D.4.2.
D = distance, mm, from the arc to the person (working distance). See Table D.4.3 for typical working distances.

If the arcing time, t, in Equation D.4.3(c) is longer than 2 seconds, consider how long a person is likely to remain in the location of the arc flash. It is likely that a person exposed to an arc flash will move away quickly if it is physically possible, and 2 seconds is a reasonable maximum time for calculations. Sound engineering judgment should be used in applying the 2-second maximum clearing time, because there could be circumstances where an employee's egress is inhibited. For example, a person in a bucket truck or a person who has crawled into equipment will need more time to move away.

Table D.4.3 Typical Working Distances

Classes of Equipment	Typical Working Distance* (mm)
15-kV switchgear	910
5-kV switchgear	910
Low-voltage switchgear	610
Low-voltage MCCs and panelboards	455
Cable	455
Other	To be determined in field

* Typical working distance is the sum of the distance between the worker and the front of the equipment and the distance from the front of the equipment to the potential arc source inside the equipment.

D.4.4 Incident Energy at Working Distance — Theoretical Equation. The following theoretically derived equation can be applied in cases where the voltage is over 15 kV or the gap is outside the range:

$$E = 2.142 \times 10^6 V I_{bf} \left(\frac{t}{D^2} \right) \quad \text{[D.4.4]}$$

where:
E = incident energy, J/cm^2
V = system voltage, kV
I_{bf} = available three-phase bolted fault current
t = arcing time, sec
D = distance (mm) from the arc to the person (working distance)

For voltages over 15 kV, arcing fault current and bolted fault current are considered equal.

D.4.5 Arc Flash Boundary. The arc flash boundary is the distance at which a person is likely to receive a second degree burn. The onset of a second degree burn is assumed to be when the skin receives 5.0 J/cm^2 of incident energy.

For the empirically derived equation,

$$D_B = \left[4.184 C_f E_n \left(\frac{t}{0.2} \right) \left(\frac{610^x}{E_B} \right) \right]^{\frac{1}{x}} \quad \text{[D.4.5(a)]}$$

For the theoretically derived equation,

$$D_B = \sqrt{2.142 \times 10^6 V I_{bf} \left(\frac{t}{E_B} \right)} \quad \text{[D.4.5(b)]}$$

where:
D_B = distance (mm) of the arc flash boundary from the arcing point
C_f = calculation factor
= 1.0 for voltages above 1 kV
= 1.5 for voltages at or below 1 kV
E_n = incident energy normalized
t = time, sec
x = distance exponent from Table D.4.2
E_B = incident energy in J/cm^2 at the distance of the arc flash boundary
V = system voltage, kV
I_{bf} = bolted three-phase available short-circuit current

Informational Note: These equations could be used to determine whether selected personal protective equipment (PPE) is adequate to prevent thermal injury at a specified distance in the event of an arc flash.

D.4.6 Current-Limiting Fuses. The formulas in this section were developed for calculating arc flash energies for use with current-limiting Class L and Class RK1 fuses. The testing was done at 600 V and at a distance of 455 mm, using commercially available fuses from one manufacturer. The following variables are noted:

I_{bf} = available three-phase bolted fault current (symmetrical rms), kA

E = incident energy, J/cm^2

(A) Class L Fuses 1601 A through 2000 A. Where $I_{bf} <$ 22.6 kA, calculate the arcing current using Equation

D.4.2(a), and use time-current curves to determine the incident energy using Equations D.4.3(a), D.4.3(b), and D.4.3(c).

Where 22.6 kA ≤ I_{bf} ≤ 65.9 kA,

$$E = 4.184\left(-0.1284 I_{bf} + 32.262\right) \quad \text{[D.4.6(a)]}$$

Where 65.9 kA < I_{bf} ≤106 kA,

$$E = 4.184\left(-0.5177 I_{bf} + 57.917\right) \quad \text{[D.4.6(b)]}$$

Where I_{bf} >106 kA, contact the manufacturer.

(B) Class L Fuses 1201 A through 1600 A. Where I_{bf} <15.7 kA, calculate the arcing current using Equation D.4.2(a), and use time-current curves to determine the incident energy using Equations D.4.3(a), D.4.3(b), and D.4.3(c).

Where 15.7 kA ≤ I_{bf} ≤ 31.8 kA,

$$E = 4.184\left(-0.1863 I_{bf} + 27.926\right) \quad \text{[D.4.6(c)]}$$

Where 44.1 kA ≤ I_{bf} ≤ 65.9 kA,

$$E = 12.3\,\text{J/cm}^2\left(2.94\ \text{cal/cm}^2\right) \quad \text{[D.4.6(e)]}$$

Where 65.9 kA <I_{bf} ≤106 kA,

$$E = 4.184\left(-0.0631 I_{bf} + 7.0878\right) \quad \text{[D.4.6(f)]}$$

Where I_{bf} >106 kA, contact the manufacturer.

(C) Class L Fuses 801 A through 1200 A. Where I_{bf} <15.7 kA, calculate the arcing current using Equation D.4.2(a), and use time-current curves to determine the incident energy per Equations D.4.3(a), D.4.3(b), and D.4.3(c).

Where 15.7 kA ≤ I_{bf} ≤ 22.6 kA,

$$E = 4.184\left(-0.1928 I_{bf} + 14.226\right) \quad \text{[D.4.6(g)]}$$

Where 22.6 kA < I_{bf} ≤ 44.1 kA,

$$E = 4.184\begin{pmatrix} 0.0143 I_{bf}^{\,2} - 1.3919 I_{bf} \\ +34.045 \end{pmatrix} \quad \text{[D.4.6(h)]}$$

Where 44.1 kA < I_{bf} ≤ 106 kA,

$$E = 1.63 \quad \text{[D.4.6(i)]}$$

Where I_{bf} >106 kA, contact the manufacturer.

(D) Class L Fuses 601 A through 800 A. Where I_{bf} <15.7 kA, calculate the arcing current using Equation D.4.2(a), and use time-current curves to determine the incident energy using Equations D.4.3(a), D.4.3(b), and D.4.3(c).

Where 15.7 kA ≤ I_{bf} ≤ 44.1 kA,

$$E = 4.184\left(-0.0601 I_{bf} + 2.8992\right) \quad \text{[D.4.6(j)]}$$

Where 44.1 kA < I_{bf} ≤106 kA,

$$E = 1.046 \quad \text{[D.4.6(k)]}$$

Where I_{bf} > 106 kA, contact the manufacturer.

(E) Class RK1 Fuses 401 A through 600 A. Where I_{bf} < 8.5 kA, calculate the arcing current using Equation D.4.2(a), and use time-current curves to determine the incident energy using Equations D.4.3(a), D.4.3(b), and D.4.3(c).

Where 8.5 kA ≤ I_{bf} ≤ 14 kA,

$$E = 4.184\left(-3.0545 I_{bf} + 43.364\right) \quad \text{[D.4.6(l)]}$$

Where 14 kA < I_{bf} ≤ 15.7 kA,

$$E = 2.510 \quad \text{[D.4.6(m)]}$$

Where 15.7 kA < I_{bf} ≤ 22.6 kA,

$$E = 4.184\left(-0.0507 I_{bf} + 1.3964\right) \quad \text{[D.4.6(n)]}$$

Where 22.6 kA < I_{bf} ≤ 106 kA,

$$E = 1.046 \quad \text{[D.4.6(o)]}$$

Where I_{bf} > 106 kA, contact the manufacturer.

(F) Class RK1 Fuses 201 A through 400 A. Where I_{bf} < 3.16 kA, calculate the arcing current using Equation D.4.2(a), and use time-current curves to determine the incident energy using Equations D.4.3(a), D.4.3(b), and D.4.3(c).

Where 3.16 kA ≤ I_{bf} ≤ 5.04 kA,

$$E = 4.184\left(-19.053 I_{bf} + 96.808\right) \quad \text{[D.4.6(p)]}$$

Where 5.04 kA < I_{bf} ≤ 22.6 kA,

$$E = 4.184\left(-0.0302 I_{bf} + 0.9321\right) \quad \text{[D.4.6(q)]}$$

Where 22.6 kA < I_{bf} ≤ 106 kA,

$$E = 1.046 \quad \text{[D.4.6(r)]}$$

Where I_{bf} >106 kA, contact the manufacturer.

(G) Class RK1 Fuses 101 A through 200 A. Where I_{bf} <1.16 kA, calculate the arcing current using Equation D.4.2(a), and use time-current curves to determine the incident energy using Equations D.4.3(a), D.4.3(b), and D.4.3(c).

Where 1.16 kA ≤ I_{bf} ≤1.6 kA,

$$E = 4.184(-18.409 I_{bf} + 36.355) \quad \textbf{[D.4.6(s)]}$$

Where 1.6 kA < I_{bf} ≤3.16 kA,

$$E = 4.184(-4.2628 I_{bf} + 13.721) \quad \textbf{[D.4.6(t)]}$$

Where 3.16 kA < I_{bf} ≤106 kA,

$$E = 1.046 \quad \textbf{[D.4.6(u)]}$$

Where I_{bf} > 106 kA, contact the manufacturer.

(H) Class RK1 Fuses 1 A through 100 A. Where I_{bf} < 0.65 kA, calculate the arcing current using Equation D.4.2(a), and use time-current curves to determine the incident energy using Equations D.4.3(a), D.4.3(b), and D.4.3(c).

Where 0.65 kA ≤ I_{bf} ≤ 1.16 kA,

$$E = 4.184(-11.176 I_{bf} + 13.565) \quad \textbf{[D.4.6(v)]}$$

Where 1.16 kA < I_{bf} ≤ 1.4 kA,

$$E = 4.184(-1.4583 I_{bf} + 2.2917) \quad \textbf{[D.4.6(w)]}$$

Where 1.4 kA < I_{bf} ≤ 106 kA,

$$E = 1.046 \quad \textbf{[D.4.6(x)]}$$

Where I_{bf} > 106 kA, contact the manufacturer.

D.4.7 Low-Voltage Circuit Breakers. The equations in Table D.4.7 can be used for systems with low-voltage circuit breakers. The results of the equations will determine the incident energy and arc flash boundary when I_{bf} is within the range as described. Time-current curves for the circuit breaker are not necessary within the appropriate range.

When the bolted fault current is below the range indicated, calculate the arcing current using Equation D.4.2(a), and use time-current curves to determine the incident energy using Equations D.4.3(a), D.4.3(b), and D.4.3(c).

Table D.4.7 Incident Energy and Arc Flash Protection Boundary by Circuit Breaker Type and Rating

			480 V and Lower		575 V–600 V	
Rating (A)	**Breaker Type**	**Trip Unit Type**	**Incident Energy (J/cm²)[a]**	**Arc Flash Boundary (mm)[a]**	**Incident Energy (J/cm²)[a]**	**Arc Flash Boundary (mm)[a]**
100–400	MCCB	TM or M	$0.189\ I_{bf} + 0.548$	$9.16\ I_{bf} + 194$	$0.271\ I_{bf} + 0.180$	$11.8\ I_{bf} + 196$
600–1200	MCCB	TM or M	$0.223\ I_{bf} + 1.590$	$8.45\ I_{bf} + 364$	$0.335\ I_{bf} + 0.380$	$11.4\ I_{bf} + 369$
600–1200	MCCB	E, LI	$0.377\ I_{bf} + 1.360$	$12.50\ I_{bf} + 428$	$0.468\ I_{bf} + 4.600$	$14.3\ I_{bf} + 568$
1600–6000	MCCB or ICCB	TM or E, LI	$0.448\ I_{bf} + 3.000$	$11.10\ I_{bf} + 696$	$0.686\ I_{bf} + 0.165$	$16.7\ I_{bf} + 606$
800–6300	LVPCB	E, LI	$0.636\ I_{bf} + 3.670$	$14.50\ I_{bf} + 786$	$0.958\ I_{bf} + 0.292$	$19.1\ I_{bf} + 864$
800–6300	LVPCB	E, LS[b]	$4.560\ I_{bf} + 27.230$	$47.20\ I_{bf} + 2660$	$6.860\ I_{bf} + 2.170$	$62.4\ I_{bf} + 2930$

MCCB: Molded-case circuit breaker.
TM: Thermal-magnetic trip units.
M: Magnetic (instantaneous only) trip units.
E: Electronic trip units have three characteristics that may be used separately or in combination: L: Long time, S: Short time, I: Instantaneous.
ICCB: Insulated-case circuit breaker.
LVPCB: Low-voltage power circuit breaker.
[a] I_{bf} is in kA; working distance is 455 mm (18 in.).
[b] Short-time delay is assumed to be set at maximum.

The range of available three-phase bolted fault currents is from 700 A to 106,000 A. Each equation is applicable for the following range:

$$I_1 < I_{bf} < I_2$$

where:

I_1 = minimum available three-phase, bolted, short-circuit current at which this method can be applied. I_1 is the lowest available three-phase, bolted, short-circuit current level that causes enough arcing current for instantaneous tripping to occur, or, for circuit breakers with no instantaneous trip, that causes short-time tripping to occur.

I_2 = interrupting rating of the circuit breaker at the voltage of interest.

To find I_1, the instantaneous trip (I_t) of the circuit breaker must be found. I_t can be determined from the time-current curve, or it can be assumed to be 10 times the rating of the circuit breaker for circuit breakers rated above 100 amperes. For circuit breakers rated 100 amperes and below, a value of I_t = 1300 A can be used. When short-time delay is utilized, I_t is the short-time pickup current.

The corresponding bolted fault current, I_{bf}, is found by solving the equation for arc current for box configurations by substituting I_t for arcing current. The 1.3 factor in Equation D.4.7(b) adjusts current to the top of the tripping band.

$$\lg(1.3I_t) = 0.084 + 0.096V + 0.586(\lg I_{bf}) + 0.559V(\lg I_{bf}) \quad \textbf{[D.4.7(a)]}$$

At 600 V,

$$\lg I_1 = 0.0281 + 1.091\lg(1.3I_t) \quad \textbf{[D.4.7(b)]}$$

At 480 V and lower,

$$\lg I_1 = 0.0407 + 1.17\ \lg(1.3I_t) \quad \textbf{[D.4.7(c)]}$$

$$I_{bf} = I_1 = 10^{\lg I_1} \quad \textbf{[D.4.7(d)]}$$

D.4.8 References. The complete data, including a spreadsheet calculator to solve the equations, can be found in IEEE 1584, *Guide for Performing Arc Flash Hazard Calculations*. IEEE publications are available from the Institute of Electrical and Electronics Engineers, 445 Hoes Lane, P.O. Box 1331, Piscataway, NJ 08855-1331, USA (http://standards.ieee.org/).

D.5 Direct-Current Incident Energy Calculations.

D.5.1 Maximum Power Method. The following method of estimating dc arc flash incident energy that follows was published in the *IEEE Transactions on Industry Applications (see reference 2, which follows)*. This method is based on the concept that the maximum power possible in a dc arc will occur when the arcing voltage is one-half the system voltage. Testing completed for Bruce Power *(see reference 3, which follows)* has shown that this calculation is conservatively high in estimating the arc flash value. This method applies to dc systems rated up to 1000 V.

$$I_{arc} = 0.5 \times I_{bf}$$
$$IE_m = 0.01 \times V_{sys} \times I_{arc} \times T_{arc} / D^2$$

where:

I_{arc} = arcing current amperes
I_{bf} = system bolted fault current amperes
IE_m = estimated dc arc flash incident energy at the maximum power point cal/cm^2
V_{sys} = system voltage volts
T_{arc} = arcing time sec
D = working distance cm

For exposures where the arc is in a box or enclosure, it would be prudent to use a multiplying factor of 3 for the resulting incident energy value.

D.5.2 Detailed Arcing Current and Energy Calculations Method. A thorough theoretical review of dc arcing current and energy was published in the *IEEE Transactions on Industry Applications*. Readers are advised to refer to that paper *(see reference 1)* for those detailed calculations.

References:

1. "DC-Arc Models and Incident-Energy Calculations," Ammerman, R.F.; et al.; *IEEE Transactions on Industry Applications*, Vol. 46, No. 5.

2. "Arc Flash Calculations for Exposures to DC Systems," Doan, D.R., *IEEE Transactions on Industry Applications*, Vol. 46, No. 6.

3. "DC Arc Hazard Assessment Phase II", Copyright Material, Kinectrics Inc., Report No. K-012623-RA-0002-R00.

D.5.3 Short Circuit Current. The determination of short circuit current is necessary in order to use Table 130.7(C)(15)(B). The arcing current is calculated at 50 percent of the dc short-circuit value. The current that a battery will deliver depends on the total impedance of the short-circuit path. A conservative approach in determining the short-circuit current that the battery will deliver at 25°C is to assume that the maximum available short-circuit current is 10 times the 1 minute ampere rating (to 1.75 volts per cell at 25°C and the specific gravity of 1.215) of the battery. A more accurate value for the short-circuit current for the specific application can be obtained from the battery manufacturer.

References:

1. IEEE 946, *Recommended Practice for the Design of DC Auxiliary Powers Systems for Generating Stations*.

Informative Annex E Electrical Safety Program

This informative annex is not a part of the requirements of this NFPA document but is included for informational purposes only.

(See 110.1, Electrical Safety Program.)

E.1 Typical Electrical Safety Program Principles. Electrical safety program principles include, but are not limited to, the following:

(1) Inspecting and evaluating the electrical equipment
(2) Maintaining the electrical equipment's insulation and enclosure integrity
(3) Planning every job and document first-time procedures
(4) De-energizing, if possible *(see 120.1)*
(5) Anticipating unexpected events
(6) Identifying the electrical hazards and reduce the associated risk
(7) Protecting employees from shock, burn, blast, and other hazards due to the working environment
(8) Using the right tools for the job
(9) Assessing people's abilities
(10) Auditing the principles

E.2 Typical Electrical Safety Program Controls. Electrical safety program controls can include, but are not limited to, the following:

(1) The employer develops programs, including training, and the employees apply them.
(2) Employees are to be trained to be qualified for working in an environment influenced by the presence of electrical energy.
(3) Procedures are to be used to identify the electrical hazards and to develop plans to eliminate those hazards or to control the associated risk for those hazards that cannot be eliminated.
(4) Every electrical conductor or circuit part is considered energized until proved otherwise.
(5) De-energizing an electrical conductor or circuit part and making it safe to work on is, in itself, a potentially hazardous task.
(6) Tasks to be performed on or near exposed energized electrical conductors and circuit parts are to be identified and categorized.
(7) Precautions appropriate to the working environment are to be determined and taken.
(8) A logical approach is to be used to determine the associated risk of each task.

E.3 Typical Electrical Safety Program Procedures. Electrical safety program procedures can include, but are not limited to determination and assessment of the following:

(1) Purpose of task
(2) Qualifications and number of employees to be involved
(3) Identification of hazards and assessment of risks of the task
(4) Limits of approach
(5) Safe work practices to be used
(6) Personal protective equipment (PPE) involved
(7) Insulating materials and tools involved
(8) Special precautionary techniques
(9) Electrical single-line diagrams
(10) Equipment details
(11) Sketches or photographs of unique features
(12) Reference data

Informative Annex F Risk Assessment Procedure

This informative annex is not a part of the requirements of this NFPA document but is included for informational purposes only.

F.1 Risk Assessment (General). This informative annex provides guidance regarding a qualitative approach for risk assessment, including risk estimation and risk evaluation, which can be helpful in determining the protective measures that are required to reduce the likelihood of injury or damage to health occurring in the circumstances under consideration. To receive the full benefit of completing the risk assessment process the relationships between the source or cause of risk and the effects of the hierarchy of controls on those causes must be understood. This informative annex is intended to provide guidance.

Risk assessment is an analytical process consisting of a number of discrete steps intended to ensure that hazards are properly identified and analyzed with regard to their severity and the likelihood of their occurrence. Once hazards have been identified and analyzed, the risk associated with those hazards can be estimated using the parameters outlined in F.2.1. Appropriate protective measures can then be implemented and evaluated in order to determine if adequate risk reduction has been achieved.

Risk assessment includes a comprehensive review of the hazards, the associated foreseeable tasks, and the protective measures that are required in order to maintain a tolerable level of risk, including the following:

(1) Identifying and analyzing electrical hazards
(2) Identifying tasks to be performed
(3) Documenting hazards associated with each task
(4) Estimating the risk for each hazard/task pair
(5) Determining the appropriate protective measures needed to adequately reduce the level of risk

Figure F.1(a) illustrates the steps to be taken and the decisions to be considered when performing an electrical work risk assessment. See 110.1 for a hazard and risk evaluation procedure. Figure F.1(b) illustrates in more detail the steps of the risk analysis, assessment, and evaluation process.

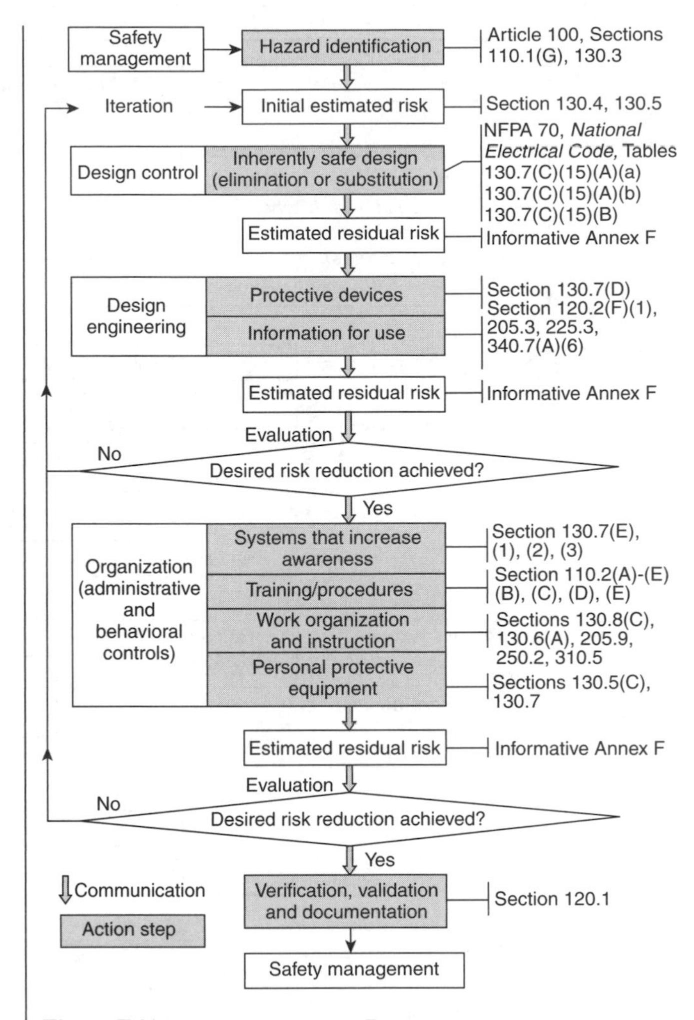

Figure F.1(a) Risk Assessment Process.

F.1.1 Responsibility. Electrical system designers, constructors, and users have responsibilities for defining and achieving tolerable risk. The supplier and the user either separately or jointly identify hazards, estimate risks, and reduce risks to a tolerable level within the scope of their respective work activities. Although the responsibilities of the supplier and the user differ over the life cycle of the electrical equipment, each entity should use the risk assessment process.

In general, the electrical system supplier is responsible for the design, construction, and information for operation and maintenance of the electrical system, while the user is responsible for the operation and maintenance of the electrical system.

Suppliers and users should involve qualified personnel in meeting their respective responsibilities. The supplier and the user should ensure compliance with the regulations and standards applicable to their work activity. This could include regulations and standards for a specific location, a specific application, or both.

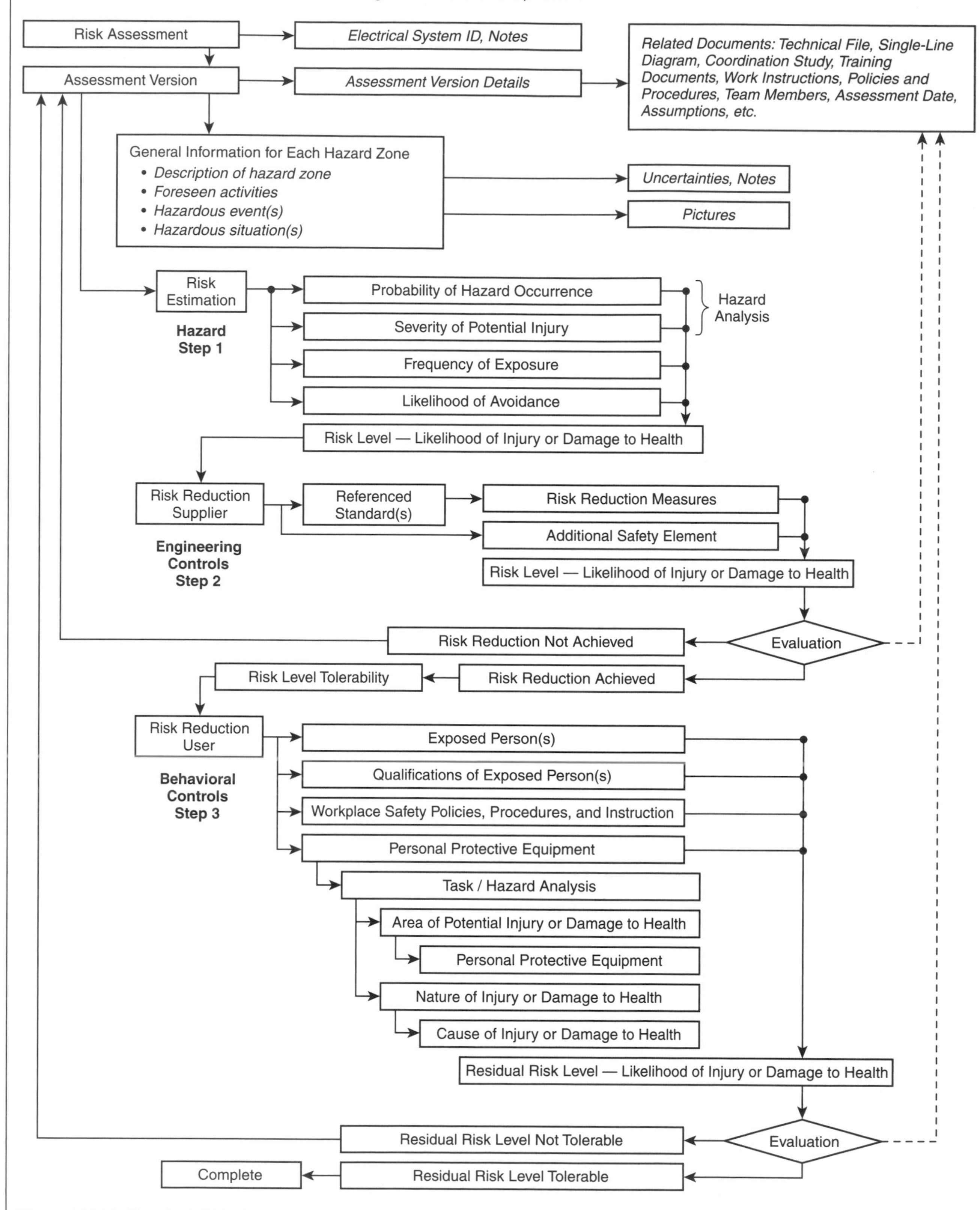

Figure F.1(b) Detailed Risk Assessment Process.

F.2 Risk Assessment.

F.2.1 Initial Risk Estimation. An initial estimation of risk should be carried out for each hazard. Risk related to the identified hazard should be derived by using the risk parameters that are shown in Figure F.2.1 including the following:

(1) Severity of injury or damage to health (Se)
(2) Likelihood of occurrence of that injury or damage to health, which is a function of all of the following:
   a. Frequency and duration of the exposure of persons to the hazard (Fr)
   b. Likelihood of occurrence of a hazardous event (Pr)
   c. Likelihood of avoiding or limiting the injury or damage to health (Av)

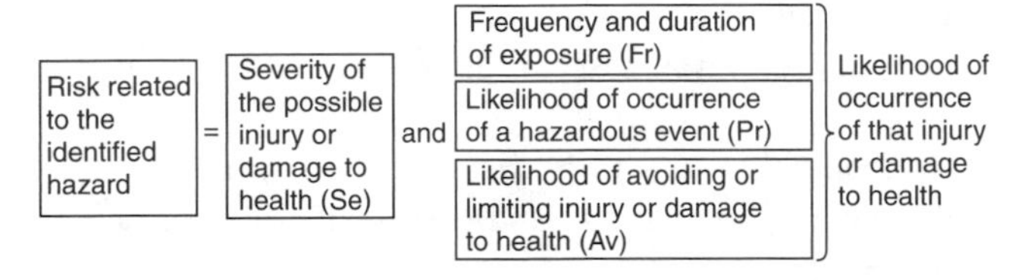

Figure F.2.1 Elements of Risk.

F.2.2 Parameters Used in Risk Estimation. In preparation for the risk assessment, parameters are estimated and can be entered into Table F.2.5. These parameters should be based on worst-case considerations for the electrical system. Different risk reduction strategies can be implemented for each hazard. The risk estimation stage is the only one at which hazards can be eliminated, thus avoiding the need for additional protective measures, such as safeguarding or complementary protective measures.

F.2.3 Severity of the Possible Injury or Damage to Health (Se). Severity of injuries or damage to health can be estimated by taking into account reversible injuries, irreversible injuries, and death. Typically, the types of hazards to be considered include, but are not limited to, shock and electrocution, burns, and impact. Choose the appropriate value of severity from Table F.2.3, based on the consequences of an injury, as follows:

(1) 8: a fatal or a significant irreversible injury, such that it will be very difficult to continue the same work after healing, if at all
(2) 6: a major or irreversible injury, in such a way that it can be possible to continue the same work after healing and can also include a severe major but reversible injury such as broken limbs
(3) 3: a reversible injury, including severe lacerations, stabbing, and severe bruises, that requires attention from a medical practitioner
(4) 1: a minor injury, including scratches and minor bruises that require attention by first aid.

Select the appropriate value for severity of the possible injury or damage to health (Se) from Table F.2.3 and insert in the Se column in Table F.2.5.

Table F.2.3 Severity of the Possible Injury or Damage to Health (Se) Classification

Severity of Injury or Damage to Health	Se Value
Irreversible — trauma, death	8
Permanent — skeletal damage, blindness, hearing loss, third degree burns	6
Reversible — minor impact, hearing damage, second degree burns	3
Reversible — minor laceration, bruises, first degree burns	1

F.2.4 Likelihood of Occurrence of Injury or Damage to Health. Each of the three parameters of likelihood of occurrence of injury or damage to health (that is, Fr, Pr, and Av) should be estimated independently of each other. A worst-case assumption needs to be used for each parameter to ensure that the protective measures, determined during risk evaluation, will provide adequate risk reduction. Generally, the use of a form of hazard/task–based evaluation is strongly recommended to ensure that proper consideration is given to the estimation of the likelihood of occurrence of injury or damage to health.

F.2.4.1 Frequency and Duration of Exposure (Fr). The following aspects should be considered to determine the level of exposure:

(1) Need for access to the hazard zone based on all modes of use; for example, normal operation and maintenance
(2) Nature of access, for example, examination, repair, and troubleshooting

It should then be possible to estimate the average interval between exposures and, thus, the average frequency of access.

This factor does not include consideration of the failure of the short-circuit interruption device(s) or the failure to use the appropriate PPE.

Select the appropriate row for frequency and duration of exposure (Fr) from Table F.2.4.1. Insert the appropriate number under the Fr column in Table F.2.5.

Table F.2.4.1 Frequency and Duration of Exposure (Fr) Classification

Frequency of Exposure	Fr Value (for Duration >10 min)
≤ 1 per hour	5
> 1 per hour to ≤ 1 per day	5
> 1 per day to ≤ 1 every 2 weeks	4
> 1 every 2 weeks to ≤ 1 per year	3
> 1 per year	2

F.2.4.2 Likelihood of Occurrence of a Hazardous Event (Pr). The occurrence of a hazardous event influences the likelihood of the occurrence of injury or damage to health. The possibility of the hazardous event occurring should describe the likelihood of the event materializing during the use or foreseeable misuse, or both, of the electrical system or process. Subjectivity may have a substantial impact on the result of the risk assessment. The use of subjective information should be minimized as far as reasonably practicable.

The likelihood of occurrence of the hazardous event should be estimated independently of other related parameters (Fr and Av) and will typically be based on the results of the completed study of the arc flash potential. The worst-case scenario should be used for this parameter to ensure that short-circuit interruption device(s) have, where practicable, been properly selected and installed and will provide adequate protection.

Elements of the electrical system that are intended to ensure an intrinsically safe design shall be taken into consideration in the determination of the likelihood of the hazardous event(s). These can include, but are not limited to, the mechanical structure, electrical devices, and electronic controls integral to the system, the process, or both at the time of the analysis. Types of components that could contribute to an inherently safe design include, but are not limited to, current-limiting devices and ground-fault circuit interrupters.

This parameter can be estimated by taking into account the following factors:

(1) The predictability of the performance of component parts of the electrical system relevant to the hazard in different modes of use (e.g., normal operation, maintenance, fault finding).

At this point in the risk assessment process, the protective effect of any personal protective equipment (PPE) and other protective measures should not be taken into account. This is necessary in order to estimate the amount of risk that will be present if the PPE and other protective measures are not in place at the time of the exposure. In general terms, it must be considered whether the electrical system being assessed has the propensity to act in an unexpected manner. The electrical system performance will vary from very predictable to not predictable. Unexpected events cannot be discounted until it can be clearly demonstrated that the electrical system will perform as expected.

Informational Note: Predictability is often linked to the complexity of the electrical system and the characteristics of the energy supply.

(2) The specified or foreseeable characteristics of human behavior with regard to interaction with the component parts of the machine relevant to the hazard, which can be characterized by one or both of the following:

a. Stress (e.g., due to time constraints, work task, perceived damage limitation)

b. Lack of awareness of information relevant to the hazard

Human behavior will be influenced by factors such as skills, training, experience, and complexity of the machine or the process.

These attributes are not usually directly under the influence of the electrical system designer, but a task analysis will reveal activities in which total awareness of all issues, including unexpected outcomes, cannot be reasonably assumed. "Very high" likelihood of occurrence of a hazardous event should be selected to reflect normal workplace constraints and worst-case considerations. Positive reasons (e.g., well-defined application and a high level of user competence) are required for any lower values to be used.

Any required or assumed skills, knowledge, and so forth, should be stated in the information for use.

Select the appropriate row for likelihood of occurrence of a hazardous event (Pr) from Table F.2.4.2.

Indicate the appropriate risk level under the Pr column in Table F.2.5.

Table F.2.4.2 Likelihood of a Hazardous Event (Pr) Classification

Likelihood of a Hazardous Event	Pr Value
Very high	5
Likely	4
Possible	3
Rare	2
Negligible	1

F.2.4.3 Likelihood of Avoiding or Limiting Injury or Damage to Health (Av). This parameter can be estimated by taking into account aspects of the electrical system design and its intended application that can help to avoid or limit the injury or damage to health from a hazard, including the following examples:

(1) Sudden or gradual appearance of the hazardous event; for example, an explosion caused by high fault values under short-circuit conditions.
(2) Spatial possibility to withdraw from the hazard.
(3) Nature of the component or system; for example, the use of touch-safe components, which reduce the likelihood of contact with energized parts. Working in close proximity to high voltage can increase the likelihood of personnel being exposed to hazards due to approach to live parts.
(4) Likelihood of recognition of a hazard; for example, as an electrical hazard, a copper bar does not change its appearance, whether it is under voltage or not. To recognize the presence of the hazard, an instrument is needed to establish whether or not electrical equipment is energized; thus, both inadvertent and intentional contact need to be considered.

Select the appropriate row for likelihood of avoiding or limiting injury or damage to health (Av) from Table F.2.4.3.

Insert the appropriate value for risk level in the Av column in Table F.2.5.

Table F.2.4.3 Likelihood of Avoiding or Limiting Injury or Damage to Health (Av) Classification

Likelihood of Avoiding or Limiting Injury or Damage to Health	Av Value
Impossible	5
Rare	3
Probable	1

F.2.5 Risk Level and Likelihood of Injury or Damage to Health. Once the parameters for each hazard under consideration have been entered in Table F.2.5, the information can be used in the first step of the risk assessment process as outlined in Figure F.1(a).

Table F.2.5 Parameters for Determining Risk Levels and Likelihood of Injury or Damage to Health (See Figure F.2.1)

Zone No.	Hazard	Se	Fr	Pr	Av

F.3 Risk Reduction.

F.3.1 Protective Measures. Once the risk prior to the application of protective measures has been estimated, all practicable efforts must be made to reduce the risk of injury or damage to health. Careful consideration of failure modes is an important part of risk reduction. Care should be taken to ensure that both technical and behavioral failures, which could result in ineffective risk reduction, are taken into account during the risk reduction stage of the risk assessment.

Situations in which hazard elimination cannot be attained typically require a balanced approach in order to reduce the likelihood of injury or damage to health. For example, the effective control of access to an electrical system requires the use of barriers, awareness placards, safe operating instructions, qualification and training, and PPE personnel protective equipment as required by this standard, as well as initial and refresher or periodic training for all affected personnel in the area. Engineering controls alone are not sufficient to reduce the remaining risk to a tolerable level. Typically, all five areas of risk reduction must be implemented to achieve the desired result.

Consideration of all five of the items in F.3.1.1 through F.3.1.5 is required to establish an adequate risk reduction strategy.

F.3.1.1 Engineering Controls. Engineering controls can have a substantial impact on risk. They should, where practicable, be considered and analyzed. Typically, engineering controls take the form of barriers and other safeguarding devices as described in *NFPA 70, National Electrical Code*; IEC 60204-1 ed 5.1 Consol. with am 1, *Safety of Machinery — Electrical Equipment of Machines — Part 1: General Requirements*; and NFPA 79, *Electrical Standard for Industrial Machinery*.

F.3.1.2 Awareness Devices. Awareness means can be used to complement the effects of engineering controls with regard to risk reduction. They should be chosen based on the design configuration for each specific application and their potential effectiveness during foreseen interaction. Each design and configuration can require unique awareness devices in order to have the desired impact on risk. Typically, awareness means take the form of signs, visual alarms, audible alarms, and so forth.

F.3.1.3 Procedures. Procedures and instructions that are required for individual(s) to safely interact with the electrical system should be identified. The procedures and instructions should include descriptions of the hazards, the possible hazardous events, hazardous situations, and the protective measures that need to be implemented. Procedures and instructions should also be used to communicate foreseeable misuse of the system that could contribute to an increased level of risk. Typically, formal procedures are provided in written form; however, in some cases, verbal instruction can be provided. Care should be taken in the latter case to ensure that the verbal instructions will have the desired impact on risk.

F.3.1.4 Training. Training, with regard to the proper interaction and for foreseeable inappropriate interaction with the electrical system, must be completed. The intent of the training is to ensure that all affected personnel are able to understand when and how hazardous situations can arise and how to best reduce the risk associated with those situations. Typically, training for individuals interacting with electrical systems will include technical information regarding hazards, hazardous situations, or both as well as information related to potential failure modes that could affect risk. This type of training generally will be provided by a trainer who has an in-depth understanding of electrical system design, as well as experience in the field of adult education. Less technical training content could be appropriate in situations in which only awareness of electrical hazards is needed to ensure that unqualified personnel do not interact with the electrical system.

F.3.1.5 Personal Protective Equipment (PPE). The electrical system must be analyzed in order to determine the appropriate PPE. Once the appropriate PPE has been determined, personnel must maintain and use it as required in order to ensure that residual risk remains at the desired level.

F.4 Risk Evaluation.

F.4.1 Risk Evaluation. Once the appropriate protective measures described in F.3.1 have been applied, the effect of those measures on the elements of risk *(see Figure F.2.1)* should be taken into account. Each type of protective measure could affect one or more of the elements that contribute to risk. The effects on risk or on the individual elements of risk, should be considered in the final risk estimation. The cumulative effect of the final combination of protective measures can then be used to estimate the residual risk. Paragraphs F.4.1.1 through F.4.1.5 provide a general, non-exhaustive outline that can be used as a guide to the final estimation of risk.

F.4.1.1 Design — Elimination or Substitution by Design.

(a) Elimination of the hazard — impacts both severity of injury or damage to health and likelihood of injury or damage to health

Failure mode(s) examples:

(1) Component(s) failure
(2) Application of an incorrect construction or manufacturing specification
(3) Incorrect calculation (that is, potential energy, toxicity, strength, durability)
(4) Inadequate procurement control
(5) Incorrect or insufficient maintenance, or both

(b) Substitution — can affect severity of injury or damage to health, frequency of exposure to the hazard under consideration, or the likelihood of avoiding or limiting injury or damage to health, depending on which method of substitution or combination thereof is applied.

Failure mode(s) examples:

(1) Unexpected or unanticipated interaction
(2) Excessive production pressure
(3) Inadequate procurement control

F.4.1.2 Design — Use of Engineering Controls.

(a) Greatest impact on the likelihood of a hazardous event(s) under certain circumstances

(b) No impact on severity of injury or damage to health

Failure mode(s) examples:

(1) Incorrect application of construction or manufacturing specification
(2) Unanticipated tasks
(3) Incentive to circumvent or reduce effectiveness
(4) Excessive production pressure
(5) Protective system failure

F.4.1.3 Use of Systems that Increase Awareness of Potential Hazards.

(a) Potential impact on avoiding or limiting injury or damage to health

(b) Potential impact on inadvertent exposure

(c) Minimal or no impact on severity of injury or damage to health

Failure mode(s) examples:

(1) Too many warning signs
(2) Depreciation of effect over time
(3) Lack of understanding

F.4.1.4 Organization and Application of a Safe System of Work.

F.4.1.4.1 Personnel training.

(a) Greatest impact on avoiding or limiting injury or damage to health

(b) Minimal, if any, impact on severity of injury or damage to health

(c) Possible impact on the likelihood of a hazardous event(s) under certain circumstances

Failure mode(s) examples:

(1) Training not understood
(2) Identified hazards not clearly communicated
(3) Depreciation of effect over time
(4) Training material not current
(5) Training not consistent with instructions
(6) Training material not inclusive of detail regarding how to perform work

F.4.1.4.2 Access restrictions.

(a) Greatest impact on exposure

(b) No impact on severity of injury or damage to health

Failure mode(s) examples:

(1) Work permit system does not exist
(2) Competency complacency
(3) Insufficient monitoring, control, or corrective actions, or combination thereof

F.4.1.4.3 Safe work procedures.

(a) Greatest impact on avoiding or limiting injury or damage to health

(b) Minimal, if any, impact on severity of injury or damage to health

(c) Possible impact on the likelihood of a hazardous event(s) under certain circumstances

Failure mode(s) examples:

(1) Inconsistent with the current culture
(2) Procedures not current or accessible
(3) Does not consider all tasks, hazards, hazardous situations, or combination thereof
(4) Insufficient monitoring, control, corrective actions, or combination thereof
(5) Instructions not consistent with training content
(6) Content too general (e.g., "Don't touch the live parts; Be careful.")

F.4.1.4.4 Policies and instructions.

(a) Greatest impact on exposure

(b) Possible impact on the likelihood of a hazardous event(s) under certain circumstances

(c) Minimal or no impact on severity of injury or damage to health

Failure mode(s) examples:

(1) Policies and instructions inconsistent
(2) Instructions not clearly communicated or accessible
(3) Insufficient monitoring, control, or corrective actions, or combination thereof
(4) Allows personnel to make the decision to work live without adequate justification

F.4.1.5 Personal Protective Equipment (PPE).

(a) Greatest impact on avoiding or limiting injury or damage to health

(b) Potential impact on inadvertent exposure

(c) Minimal impact on severity of injury or damage to health

(d) No impact on the likelihood of a hazardous event(s)

Failure mode(s) examples:

(1) Reason for use not understood
(2) Creates barriers to effective completion of the work
(3) PPE specification inappropriate for the considered hazards
(4) Production pressure does not afford time to use or maintain
(5) Worker forgets to use when needed
(6) Excessive discomfort
(7) Perceived invulnerability
(8) Insufficient monitoring, control, corrective actions, or combination thereof

F.5 Risk Reduction Verification.

F.5.1 Verification. Once the assessment has been completed and protective measures have been determined, it is imperative to ensure that the protective measures are implemented prior to initiating the electrical work. While this procedure might not result in a reduction of the PPE required, it could improve the understanding of the properties of the hazards associated with a task to a greater extent and thus allow for improvement in the implementation of the protective measures that have been selected.

F.5.2 Auditing. For each activity that has been assessed, it might be necessary to audit the risk reduction strategy that is applicable. If an audit is required, the auditing process should take place prior to commencement of work on electrical systems. An example of a nonexhaustive audit is shown in Figure F.5.2. Each audit process might need to be specific to the properties of the electrical system, to the task to be performed, or to both.

Hazard (situation)	Risk Reduction Strategy	Confirmation (in place) Yes / No
Human factors (mistakes)	Training and instructions include details regarding hazardous situations that could arise.	
Human factors (willful disregard)	Policies and supervision are in place in order to ensure that instructions are followed.	
Unqualified person performing electrical work	Work permit system is in place to control personnel activities.	
Inappropriate overcurrent protection	Instructions include details regarding the selection or replacement of fuses and/or circuit breakers.	
Short circuits between test leads	Training and instructions include details regarding care and inspection of testing equipment.	
Meter malfunctions	Training and instructions include details regarding care and inspection of testing equipment.	
Meter misapplication	Training and instructions include details regarding use of testing equipment.	
Qualified person performing electrical work that exceeds individual's qualification	Work permit system is in place to control personnel activities.	

Figure F.5.2 Sample Auditing Form.

Informative Annex G Sample Lockout/Tagout Procedure

This informative annex is not a part of the requirements of this NFPA document but is included for informational purposes only.

Lockout is the preferred method of controlling personnel exposure to electrical energy hazards. Tagout is an alternative method that is available to employers. To assist employers in developing a procedure that meets the requirement of 120.2 of *NFPA 70E*, the sample procedure that follows is provided for use in lockout and tagout programs. This procedure can be used for a simple lockout/tagout, or as part of a complex lockout/tagout. A more comprehensive plan will need to be developed, documented, and used for the complex lockout/tagout.

LOCKOUT/TAGOUT PROCEDURE
FOR [COMPANY NAME]
OR
TAGOUT PROCEDURE FOR [COMPANY NAME]

1.0 Purpose. This procedure establishes the minimum requirements for lockout/tagout of electrical energy sources. It is to be used to ensure that conductors and circuit parts are disconnected from sources of electrical energy, locked (tagged), and tested before work begins where employees could be exposed to dangerous conditions. Sources of stored energy, such as capacitors or springs, shall be relieved of their energy, and a mechanism shall be engaged to prevent the reaccumulation of energy.

2.0 Responsibility. All employees shall be instructed in the safety significance of the lockout/tagout procedure. All new or transferred employees and all other persons whose work operations are or might be in the area shall be instructed in the purpose and use of this procedure. *[Name(s) of the person(s) or the job title(s) of the employee(s) with responsibility]* shall ensure that appropriate personnel receive instructions on their roles and responsibilities. All persons installing a lockout/tagout device shall sign their names and the date on the tag *[or state how the name of the individual or person in charge will be available]*.

3.0 Preparation for Lockout/Tagout.

3.1 Review current diagrammatic drawings (or their equivalent), tags, labels, and signs to identify and locate all disconnecting means to determine that power is interrupted by a physical break and not de-energized by a circuit interlock. Make a list of disconnecting means to be locked (tagged).

3.2 Review disconnecting means to determine adequacy of their interrupting ability. Determine if it will be possible to verify a visible open point, or if other precautions will be necessary.

3.3 Review other work activity to identify where and how other personnel might be exposed to electrical hazards. Review other energy sources in the physical area to determine employee exposure to those sources of other types of energy. Establish energy control methods for control of other hazardous energy sources in the area.

3.4 Provide an adequately rated test instrument to test each phase conductor or circuit part to verify that they are de-energized *(see Section 11.3)*. Provide a method to determine that the test instrument is operating satisfactorily.

3.5 Where the possibility of induced voltages or stored electrical energy exists, call for grounding the phase conductors or circuit parts before touching them. Where it could be reasonably anticipated that contact with other exposed energized conductors or circuit parts is possible, call for applying ground connecting devices.

4.0 Simple Lockout/Tagout. The simple lockout/tagout procedure will involve 1.0 through 3.0, 5.0 through 9.0, and 11.0 through 13.0.

5.0 Sequence of Lockout/Tagout System Procedures.

5.1 The employees shall be notified that a lockout/tagout system is going to be implemented and the reason for it. The qualified employee implementing the lockout/tagout shall know the disconnecting means location for all sources of electrical energy and the location of all sources of stored energy. The qualified person shall be knowledgeable of hazards associated with electrical energy.

5.2 If the electrical supply is energized, the qualified person shall de-energize and disconnect the electric supply and relieve all stored energy.

5.3 Wherever possible, the blades of disconnecting devices should be visually verified to be fully opened, or draw-out type circuit breakers should be verified to be completely withdrawn to the fully disconnected position.

5.4 Lockout/tagout all disconnecting means with lockout/tagout devices.

Informational Note: For tagout, one additional safety measure must be employed, such as opening, blocking, or removing an additional circuit element.

5.5 Attempt to operate the disconnecting means to determine that operation is prohibited.

5.6 A test instrument shall be used. *(See 11.3.)* Inspect the instrument for visible damage. Do not proceed if there is an indication of damage to the instrument until an undamaged device is available.

5.7 Verify proper instrument operation and then test for absence of voltage.

5.8 Verify proper instrument operation after testing for absence of voltage.

5.9 Where required, install a grounding equipment/conductor device on the phase conductors or circuit parts, to eliminate induced voltage or stored energy, before touching them. Where it has been determined that contact with other exposed energized conductors or circuit parts is possible, apply ground connecting devices rated for the available fault duty.

5.10 The equipment, electrical source, or both are now locked out (tagged out).

6.0 Restoring the Equipment, Electrical Supply, or Both to Normal Condition.

6.1 After the job or task is complete, visually verify that the job or task is complete.

6.2 Remove all tools, equipment, and unused materials and perform appropriate housekeeping.

6.3 Remove all grounding equipment/conductors/devices.

6.4 Notify all personnel involved with the job or task that the lockout/tagout is complete, that the electrical supply is being restored, and that they are to remain clear of the equipment and electrical supply.

6.5 Perform any quality control tests or checks on the repaired or replaced equipment, electrical supply, or both.

6.6 Remove lockout/tagout devices. The person who installed the devices is to remove them.

6.7 Notify the owner of the equipment, electrical supply, or both, that the equipment, electrical supply, or both are ready to be returned to normal operation.

6.8 Return the disconnecting means to their normal condition.

7.0 Procedure Involving More Than One Person. For a simple lockout/tagout and where more than one person is involved in the job or task, each person shall install his or her own personal lockout/tagout device.

8.0 Procedure Involving More Than One Shift. When the lockout/tagout extends for more than one day, it shall be verified that the lockout/tagout is still in place at the beginning of the next day. When the lockout/tagout is continued on successive shifts, the lockout/tagout is considered to be a complex lockout/tagout.

For a complex lockout/tagout, the person in charge shall identify the method for transfer of the lockout/tagout and of communication with all employees.

9.0 Complex Lockout/Tagout. A complex lockout/tagout plan is required where one or more of the following exist:

(1) Multiple energy sources (more than one)
(2) Multiple crews
(3) Multiple crafts
(4) Multiple locations
(5) Multiple employers
(6) Unique disconnecting means
(7) Complex or particular switching sequences
(8) Lockout/tagout for more than one shift; that is, new shift workers

9.1 All complex lockout/tagout procedures shall require a written plan of execution. The plan shall include the requirements in 1.0 through 3.0, 5.0, 6.0, and 8.0 through 12.0.

9.2 A person in charge shall be involved with a complex lockout/tagout procedure. The person in charge shall be at the procedure location.

9.3 The person in charge shall develop a written plan of execution and communicate that plan to all persons engaged in the job or task. The person in charge shall be held accountable for safe execution of the complex lockout/tagout plan. The complex lockout/tagout plan must address all the concerns of employees who might be exposed, and they must understand how electrical energy is controlled. The person in charge shall ensure that each person understands the electrical hazards to which they are exposed and the safety-related work practices they are to use.

9.4 All complex lockout/tagout plans identify the method to account for all persons who might be exposed to electrical hazards in the course of the lockout/tagout.

One of the following methods is to be used:

(1) Each individual shall install his or her own personal lockout or tagout device.

(2) The person in charge shall lock his/her key in a lock box.
(3) The person in charge shall maintain a sign-in/sign-out log for all personnel entering the area.
(4) Another equally effective methodology shall be used.

9.5 The person in charge can install locks/tags or direct their installation on behalf of other employees.

9.6 The person in charge can remove locks/tags or direct their removal on behalf of other employees, only after all personnel are accounted for and ensured to be clear of potential electrical hazards.

9.7 Where the complex lockout/tagout is continued on successive shifts, the person in charge shall identify the method for transfer of the lockout and the method of communication with all employees.

10.0 Discipline.

10.1 Knowingly violating this procedure will result in [*state disciplinary actions that will be taken*].

10.2 Knowingly operating a disconnecting means with an installed lockout device (tagout device) will result in [*state disciplinary actions to be taken*].

11.0 Equipment.

11.1 Locks shall be [*state type and model of selected locks*].

11.2 Tags shall be [*state type and model to be used*].

11.3 The test instrument(s) to be used shall be [*state type and model*].

12.0 Review. This procedure was last reviewed on [date] and is scheduled to be reviewed again on [date] (not more than 1 year from the last review).

13.0 Lockout/Tagout Training. Recommended training can include, but is not limited to, the following:

(1) Recognition of lockout/tagout devices
(2) Installation of lockout/tagout devices
(3) Duty of employer in writing procedures
(4) Duty of employee in executing procedures
(5) Duty of person in charge
(6) Authorized and unauthorized removal of locks/tags
(7) Enforcement of execution of lockout/tagout procedures
(8) Simple lockout/tagout
(9) Complex lockout/tagout
(10) Use of single-line and diagrammatic drawings to identify sources of energy
(11) Alerting techniques
(12) Release of stored energy
(13) Personnel accounting methods
(14) Temporary protective grounding equipment needs and requirements
(15) Safe use of test instruments

Informative Annex H Guidance on Selection of Protective Clothing and Other Personal Protective Equipment (PPE)

This informative annex is not a part of the requirements of this NFPA document but is included for informational purposes only.

H.1 Arc-Rated Clothing and Other Personal Protective Equipment (PPE) for Use with Arc Flash PPE Categories. Table 130.7(C)(15)(A)(a), Table 130.7(C)(15)(A)(b), Table 130.7(C)(15)(B), and Table 130.7(C)(16) provide guidance for the selection and use of PPE when using arc flash PPE categories.

H.2 Simplified Two-Category Clothing Approach for Use with Table 130.7(C)(15)(A)(a), Table 130.7(C)(15)(A)(b), Table 130.7(C)(15)(B), and Table 130.7(C)(16). The use of Table H.2 is a simplified approach to provide minimum PPE for electrical workers within facilities with large and diverse electrical systems. The clothing listed in Table H.2 fulfills the minimum arc-rated clothing requirements of Table 130.7(C)(15)(A)(a), Table 130.7(C)(15)(A)(b), Table 130.7(C)(15)(B), and Table 130.7(C)(16). The clothing systems listed in this table should be used with the other PPE appropriate for the arc flash PPE category *[see Table 130.7(C)(16)]*. The notes to Table 130.7(C)(15)(A)(a), Table 130.7(C)(15)(A)(b), and Table 130.7(C)(15)(B), must apply as shown in those tables.

H.3 Arc-Rated Clothing and Other Personal Protective Equipment (PPE) for Use with Risk Assessment of Electrical Hazards. Table H.3(a) provides a summary of specific sections within the *NFPA 70E* standard describing PPE for electrical hazards. Table H.3(b) provides guidance on the selection of arc-rated and other PPE for users who determine the incident energy exposure (in cal/cm^2).

Table H.2 Simplified Two-Category, Arc-Rated Clothing System

Clothing[a]	Applicable Tasks
Everyday Work Clothing Arc-rated long-sleeve shirt with arc-rated pants (minimum arc rating of 8) *or* Arc-rated coveralls (minimum arc rating of 8)	All arc flash PPE category 1 and arc flash PPE category 2 tasks listed in Table 130.7(C)(15)(A)(a), Table 130.7(C)(15)(A)(b), and Table 130.7(C)(15)(B)[b]
Arc Flash Suit A total clothing system consisting of arc-rated shirt and pants and/or arc-rated coveralls and/or arc flash coat and pants (clothing system minimum arc rating of 40)	All arc flash PPE category 3 and arc flash PPE category 4 tasks listed in Table 130.7(C)(15)(A)(a), Table 130.7(C)(15)(A)(b), and Table 130.7(C)(15)(B)[b]

[a]Note that other PPE listed in Table 130.7(C)(16), which include arc-rated face shields or arc flash suit hoods, arc-rated hard hat liners, safety glasses or safety goggles, hard hats, hearing protection, heavy-duty leather gloves, rubber insulating gloves, and leather protectors, could be required. The arc rating for a garment is expressed in cal/cm^2.

[b]The estimated available short-circuit current capacities and fault clearing times or arcing durations are listed in the text of Table 130.7(C)(15)(A)(b) and Table 130.7(C)(15)(B). Various tasks are listed in Table 130.7(C)(15)(A)(a). For tasks not listed or for power systems with greater than the estimated available short-circuit capacity or with longer than the assumed fault clearing times or arcing durations, an arc flash risk assessment is required in accordance with 130.5.

Tables H.3(a) and (b) were revised by a tentative interim amendment (TIA). See page 1.

Table H.3(a) Summary of Specific Sections Describing PPE for Electrical Hazards

Shock Hazard PPE	Applicable Section(s)
Rubber insulating gloves and leather protectors (unless the requirements of ASTM F 496 are met)	130.7(C)(7)(a)
Rubber insulating sleeves (as needed)	130.7(C)(7)(a)
Class G or E hard hat (as needed)	130.7(C)(3)
Safety glasses or goggles (as needed)	130.7(C)(4)
Dielectric overshoes (as needed)	130.7(C)(8)

Table H.3(a) Summary of Specific Sections Describing PPE for Electrical Hazards

Arc Flash Hazard PPE	Applicable Section(s)
Incident energy exposures up to 1.2 cal/cm^2	
Clothing: nonmelting or untreated natural fiber long-sleeve shirt and long pants or coverall	130.7(C)(1); 130.7(C)(9)(d)
Gloves: heavy-duty leather	130.7(C)(7)(b); 130.7(C)(10)(d)
Hard hat: class G or E	130.7(C)(3)
Face shield: covers the face, neck, and chin (as needed)	130.7(C)(3)
Safety glasses or goggles	130.7(C)(4); 130.7(C)(10)(c)
Hearing protection	130.7(C)(5)
Footwear: heavy-duty leather (as needed)	130.7(C)(10)(e)
Incident Energy Exposures ≥ 1.2 cal/cm^2	
Clothing: arc-rated clothing system with an arc rating appropriate to the anticipated incident energy exposure	130.7(C)(1); 130.7(C)(2); 130.7(C)(6); 130.7(C)(9)(d)
Clothing underlayers (when used): arc-rated or nonmelting untreated natural fiber	130.7(C)(9)(c); 130.7(C)(11); 130.7(C)(12)
Gloves:	130.7(C)(7)(b); 130.7(C)(10)(d)
Exposures ≥ 1.2 cal/cm^2 and ≤ 8 cal/cm^2: heavy-duty leather gloves	
Exposures > 8 cal/cm^2: rubber insulating gloves with their leather protectors; or arc-rated gloves	
Hard hat: class G or E	130.7(C)(1); 130.7(C)(3)
Face shield:	130.7(C)(1); 130.7(C)(3); 130.7(C)(10)(a); 130.7(C)(10)(b); 130.7(C)(10)(c)
Exposures ≥1.2 cal/cm^2 and 12 cal/cm^2: arc-rated face shield that covers the face, neck, and chin and an arc-rated balaclava or an arc-rated arc flash suit hood	
Exposures >12 cal/cm^2: arc-rated arc flash suit hood	
Safety glasses or goggles	130.7(C)(4); 130.7(C)(10)(c)
Hearing protection	130.7(C)(5)
Footwear:	130.7(C)(10)(e)
Exposures ≤4 cal/cm^2: heavy-duty leather footwear (as needed)	
Exposures > 4 cal/cm^2: heavy-duty leather footwear	

Table H.3(b) Guidance on Selection of Arc-Rated Clothing and Other PPE for Use When Incident Energy Exposure Is Determined

Incident Energy Exposure	Protective Clothing and PPE
≤ 1.2 cal/cm^2	
Protective clothing, nonmelting (in accordance with ASTM F 1506) or untreated natural fiber	Shirt (long sleeve) and pants (long) or coverall
Other PPE	Face shield for projectile protection (AN) Safety glasses or safety goggles (SR) Hearing protection Heavy-duty leather gloves or rubber insulating gloves with leather protectors (AN)
> 1.2 to 12 cal/cm^2 Arc-rated clothing and equipment with an arc rating equal to or greater than the determined incident energy *(See Note 3.)*	Arc-rated long-sleeve shirt and arc-rated pants or arc-rated coverall or arc flash suit (SR) *(See Note 3.)* Arc-rated face shield and arc-rated balaclava or arc flash suit hood (SR) *(See Note 1.)* Arc-rated jacket, parka, or rainwear (AN)
Other PPE	Hard hat Arc-rated hard hat liner (AN) Safety glasses or safety goggles (SR) Hearing protection Heavy-duty leather gloves or rubber insulating gloves with leather protectors (SR) *(See Note 4.)* Leather footwear
> 12 cal/cm^2 Arc-rated clothing and equipment with an arc rating equal to or greater than the determined incident energy *(See Note 3.)*	Arc-rated long-sleeve shirt and arc-rated pants or arc-rated coverall and/or arc flash suit (SR) Arc-rated arc flash suit hood Arc-rated gloves Arc-rated jacket, parka, or rainwear (AN)
Other PPE	Hard hat Arc-rated hard hat liner (AN) Safety glasses or safety goggles (SR) Hearing protection Arc-rated gloves or rubber insulating gloves with leather protectors (SR) *(See Note 4.)* Leather footwear

AN: As needed [in addition to the protective clothing and PPE required by 130.5(C)(1)].
SR: Selection of one in group is required by 130.5(C)(1).
Notes:
(1) Face shields with a wrap-around guarding to protect the face, chin, forehead, ears, and neck area are required by 130.7(C)(10)(c). For full head and neck protection, use a balaclava or an arc flash hood.
(2) All items not designated "AN" are required by 130.7(C).
(3) Arc ratings can be for a single layer, such as an arc-rated shirt and pants or a coverall, or for an arc flash suit or a multi-layer system consisting of a combination of arc-rated shirt and pants, coverall, and arc flash suit.
(4) Rubber insulating gloves with leather protectors provide arc flash protection in addition to shock protection. Higher class rubber insulating gloves with leather protectors, due to their increased material thickness, provide increased arc flash protection.

Informative Annex I Job Briefing and Planning Checklist

This informative annex is not a part of the requirements of this NFPA document but is included for informational purposes only.

I.1 Job Briefing and Planning Checklist. Figure I.1 illustrates considerations for a job briefing and planning checklist.

Identify
- ❑ Hazards
- ❑ Voltage levels involved
- ❑ Skills required
- ❑ Any "foreign" (secondary source) voltage source
- ❑ Any unusual work conditions
- ❑ Number of people needed to do the job
- ❑ Shock protection boundaries
- ❑ Available incident energy
- ❑ Potential for arc flash (Conduct an arc flash hazard analysis.)
- ❑ Arc flash boundary

Ask
- ❑ Can the equipment be de-energized?
- ❑ Are backfeeds of the circuits to be worked on possible?
- ❑ Is a standby person required?

Check
- ❑ Job plans
- ❑ Single-line diagrams and vendor prints
- ❑ Status board
- ❑ Information on plant and vendor resources is up to date
- ❑ Safety procedures
- ❑ Vendor information
- ❑ Individuals are familiar with the facility

Know
- ❑ What the job is
- ❑ Who else needs to know — Communicate!
- ❑ Who is in charge

Think
- ❑ About the unexpected event . . . What if?
- ❑ Lock — Tag — Test — Try
- ❑ Test for voltage — FIRST
- ❑ Use the right tools and equipment, including PPE
- ❑ Install and remove temporary protective grounding equipment
- ❑ Install barriers and barricades
- ❑ What else . . . ?

Prepare for an emergency
- ❑ Is the standby person CPR trained?
- ❑ Is the required emergency equipment available? Where is it?
- ❑ Where is the nearest telephone?
- ❑ Where is the fire alarm?
- ❑ Is confined space rescue available?
- ❑ What is the exact work location?
- ❑ How is the equipment shut off in an emergency?
- ❑ Are the emergency telephone numbers known?
- ❑ Where is the fire extinguisher?
- ❑ Are radio communications available?

Figure I.1 Sample Job Briefing and Planning Checklist.

Informative Annex J Energized Electrical Work Permit

This informative annex is not a part of the requirements of this NFPA document but is included for informational purposes only.

J.1 Energized Electrical Work Permit Sample. Figure J.1 illustrates considerations for an energized electrical work permit.

J.2 Energized Electrical Work Permit. Figure J.2 illustrates items to consider when determining the need for an energized electrical work permit.

ENERGIZED ELECTRICAL WORK PERMIT

PART I: TO BE COMPLETED BY THE REQUESTER:

Job/Work Order Number ________________

(1) Description of circuit/equipment/job location: ________________

(2) Description of work to be done: ________________

(3) Justification of why the circuit/equipment cannot be de-energized or the work deferred until the next scheduled outage:

________________ ________________

Requester/Title Date

PART II: TO BE COMPLETED BY THE ELECTRICALLY QUALIFIED PERSONS *DOING* THE WORK:

Check when complete

(1) Detailed job description procedure to be used in performing the above detailed work: ☐

(2) Description of the safe work practices to be employed: ________________ ☐

(3) Results of the shock risk assessment: ________________

(a) Voltage to which personnel will be exposed ☐

(b) Limited approach boundary ☐

(c) Restricted approach boundary ☐

(d) Necessary shock, personal, and other protective equipment to safely perform assigned task ☐

(4) Results of the arc flash risk assessment: ________________

(a) Available incident energy at the working distance or arc flash PPE category ☐

(b) Necessary arc flash personal and other protective equipment to safely perform the assigned task ☐

(c) Arc flash boundary ☐

(5) Means employed to restrict the access of unqualified persons from the work area: ________________ ☐

(6) Evidence of completion of a job briefing, including discussion of any job-related hazards: ________________ ☐

(7) Do you agree the above-described work can be done safely? ❑ Yes ❑ No (If *no*, return to requester.)

Electrically Qualified Person(s) Date

Electrically Qualified Person(s) Date

PART III: APPROVAL(S) TO PERFORM THE WORK WHILE ELECTRICALLY ENERGIZED:

Manufacturing Manager | Maintenance/Engineering Manager

Safety Manager | Electrically Knowledgeable Person

General Manager | Date

Note: Once the work is complete, forward this form to the site Safety Department for review and retention.

 NFPA 70E

Figure J.1 Sample Permit for Energized Electrical Work.

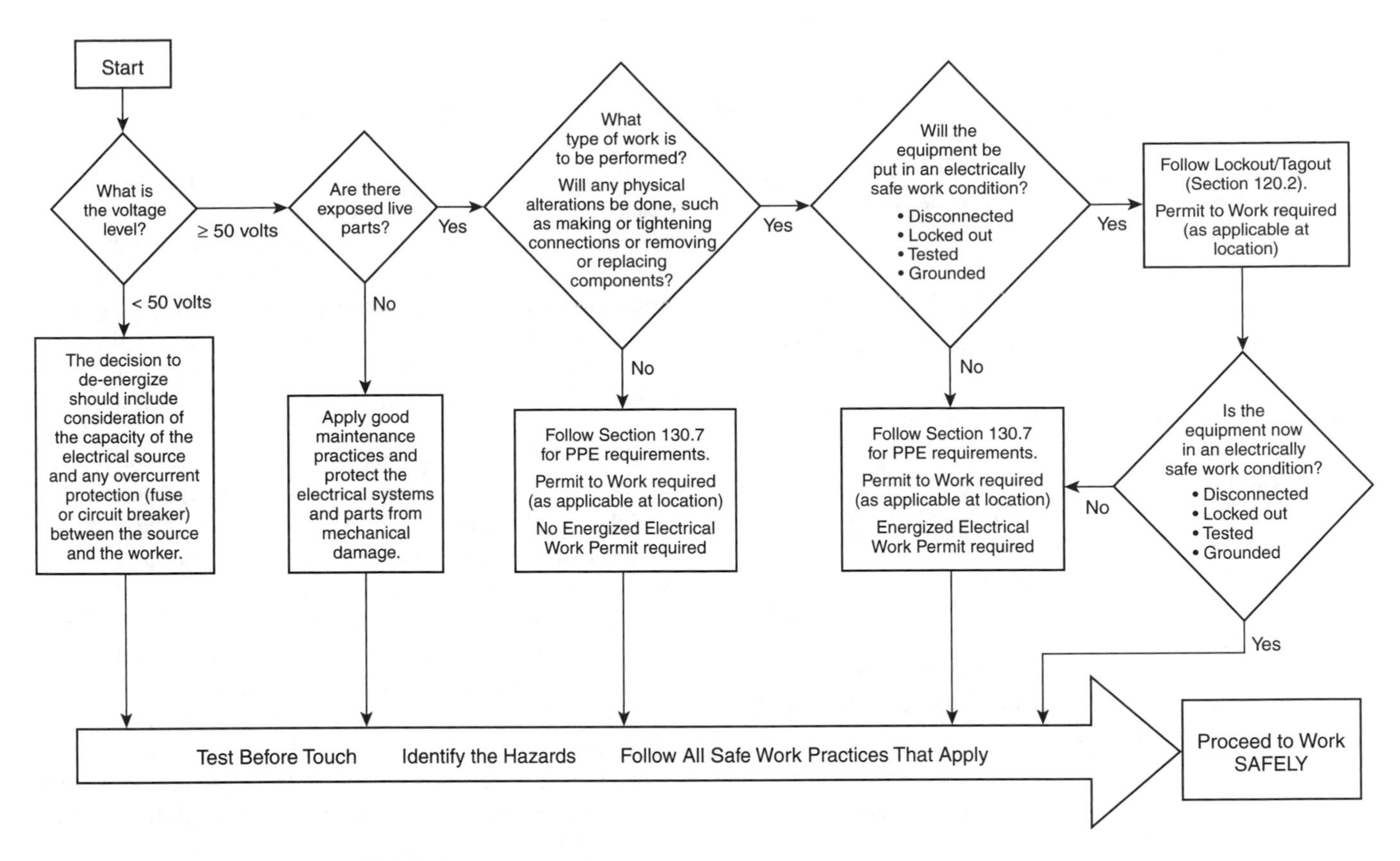

Figure J.2 Energized Electrical Work Permit Flow Chart.

Informative Annex K General Categories of Electrical Hazards

This informative annex is not a part of the requirements of this NFPA document but is included for informational purposes only.

K.1 General Categories. There are three general categories of electrical hazards: electrical shock, arc flash, and arc blast.

K.2 Electric Shock. Approximately 30,000 nonfatal electrical shock accidents occur each year. The National Safety Council estimates that about 1000 fatalities each year are due to electrocution, more than half of them while servicing energized systems of less than 600 volts.

Electrocution is the fourth leading cause of industrial fatalities, after traffic, homicide, and construction accidents. The current required to light a 7½-watt, 120-volt lamp, if passed across the chest, is enough to cause a fatality. The most damaging paths through the body are through the lungs, heart, and brain.

K.3 Arc Flash. When an electric current passes through air between ungrounded conductors or between ungrounded conductors and grounded conductors, the temperatures can reach 35,000°F. Exposure to these extreme temperatures both burns the skin directly and causes ignition of clothing, which adds to the burn injury. The majority of hospital admissions due to electrical accidents are from arc flash burns, not from shocks. Each year more than 2000 people are admitted to burn centers with severe arc flash burns. Arc flashes can and do kill at distances of 3 m (10 ft).

K.4 Arc Blast. The tremendous temperatures of the arc cause the explosive expansion of both the surrounding air and the metal in the arc path. For example, copper expands by a factor of 67,000 times when it turns from a solid to a vapor. The danger associated with this expansion is one of high pressures, sound, and shrapnel. The high pressures can easily exceed hundreds or even thousands of pounds per square foot, knocking workers off ladders, rupturing eardrums, and collapsing lungs. The sounds associated with these pressures can exceed 160 dB. Finally, material and molten metal are expelled away from the arc at speeds exceeding 1120 km/hr (700 mph), fast enough for shrapnel to completely penetrate the human body.

Informative Annex L Typical Application of Safeguards in the Cell Line Working Zone

This informative annex is not a part of the requirements of this NFPA document but is included for informational purposes only.

L.1 Application of Safeguards. This informative annex permits a typical application of safeguards in electrolytic areas where electrical hazards exist. Take, for example, an employee working on an energized cell. The employee uses manual contact to make adjustments and repairs. Consequently, the exposed energized cell and grounded metal floor could present an electrical hazard. Safeguards for this employee can be provided in the following ways:

(1) Protective boots can be worn that isolate the employee's feet from the floor and that provide a safeguard from the electrical hazard.
(2) Protective gloves can be worn that isolate the employee's hands from the energized cell and that provide a safeguard.
(3) If the work task causes severe deterioration, wear, or damage to personal protective equipment (PPE), the employee might have to wear both protective gloves and boots.
(4) A permanent or temporary insulating surface can be provided for the employee to stand on to provide a safeguard.
(5) The design of the installation can be modified to provide a conductive surface for the employee to stand on. If the conductive surface is bonded to the cell, a safeguard will be provided by voltage equalization.
(6) Safe work practices can provide safeguards. If protective boots are worn, the employee should not make long reaches over energized (or grounded) surfaces such that his or her elbow bypasses the safeguard. If such movements are required, protective sleeves, protective mats, or special tools should be used. Training on the nature of electrical hazards and proper use and condition of safeguards is, in itself, a safeguard.
(7) The energized cell can be temporarily bonded to ground.

L.2 Electrical Power Receptacles. Power supply circuits and receptacles in the cell line area for portable electric equipment should meet the requirements of 668.21 of *NFPA 70, National Electrical Code*. However, it is recommended that receptacles for portable electric equipment not be installed in electrolytic cell areas and that only pneumatic-powered portable tools and equipment be used.

Informative Annex M Layering of Protective Clothing and Total System Arc Rating

This informative annex is not a part of the requirements of this NFPA document but is included for informational purposes only.

M.1 Layering of Protective Clothing.

M.1.1 Layering of arc-rated clothing is an effective approach to achieving the required arc rating. The use of all arc-rated clothing layers will result in achieving the required arc rating with the lowest number of layers and lowest clothing system weight. Garments that are not arc-rated should not be used to increase the arc rating of a garment or of a clothing system.

M.1.2 The total system of protective clothing can be selected to take credit for the protection provided by all the layers of clothing that are worn. For example, to achieve an arc rating of 40 cal/cm^2, an arc flash suit with an arc rating of 40 cal/cm^2 could be worn over a cotton shirt and cotton pants. Alternatively, an arc flash suit with a 25 cal/cm^2 arc rating could be worn over an arc-rated shirt and arc-rated pants with an arc rating of 8 cal/cm^2 to achieve a total system arc rating of 40 cal/cm^2. This latter approach provides the required arc rating at a lower weight and with fewer total layers of fabric and, consequently, would provide the required protection with a higher level of worker comfort.

M.2 Layering Using Arc-Rated Clothing over Natural Fiber Clothing Underlayers.

M.2.1 Under some exposure conditions, natural fiber underlayers can ignite even when they are worn under arc-rated clothing.

M.2.2 If the arc flash exposure is sufficient to break open all the arc-rated clothing outerlayer or underlayers, the natural fiber underlayer can ignite and cause more severe burn injuries to an expanded area of the body. This is due to the natural fiber underlayers burning onto areas of the worker's body that were not exposed by the arc flash event. This can occur when the natural fiber underlayer continues to burn underneath arc-rated clothing layers even in areas in which the arc-rated clothing layer or layers are not broken open due to a "chimney effect."

M.3 Total System Arc Rating.

M.3.1 The total system arc rating is the arc rating obtained when all clothing layers worn by a worker are tested as a multilayer test sample. An example of a clothing system is an arc-rated coverall worn over an arc-rated shirt and arc-rated pants in which all of the garments are constructed from the same arc-rated fabric. For this two-layer arc-rated clothing system, the arc rating would typically be more than three times higher than the arc ratings of the individual layers; that is, if the arc ratings of the arc-rated coverall, shirt, and pants were all in the range of 5 cal/cm^2 to 6 cal/cm^2, the total two-layer system arc rating would be over 20 cal/cm^2.

M.3.2 It is important to understand that the total system arc rating cannot be determined by adding the arc ratings of the individual layers. In a few cases, it has been observed that the total system arc rating actually decreased when another arc-rated layer of a specific type was added to the system as the outermost layer. The only way to determine the total system arc rating is to conduct a multilayer arc test on the combination of all of the layers assembled as they would be worn.

Informative Annex N Example Industrial Procedures and Policies for Working Near Overhead Electrical Lines and Equipment

This informative annex is not a part of the requirements of this NFPA document but is included for informational purposes only.

N.1 Introduction. This informative annex is an example of an industrial procedure for working near overhead electrical systems. Areas covered include operations that could expose employees or equipment to contact with overhead electrical systems.

When working near electrical lines or equipment, avoid direct or indirect contact. Direct contact is contact with any part of the body. Indirect contact is when part of the body touches or is in dangerous proximity to any object in contact with energized electrical equipment. The following two assumptions should always be made:

(1) Lines are "live" (energized).
(2) Lines are operating at high voltage (over 1000 volts).

As the voltage increases, the minimum working clearances increase. Through arc-over, injuries or fatalities could occur, even if actual contact with high-voltage lines or equipment is not made. Potential for arc-over increases as the voltage increases.

N.2 Overhead Power Line Policy (OPP). This informative annex applies to all overhead conductors, regardless of voltage, and requires the following:

(1) That employees not place themselves in close proximity to overhead power lines. "Close proximity" is within a distance of 3 m (10 ft) for systems up to 50 kV, and should be increased 100 mm (4 in.) for every 10 kV above 50 kV.
(2) That employees be informed of the hazards and precautions when working near overhead lines.
(3) That warning decals be posted on cranes and similar equipment regarding the minimum clearance of 3 m (10 ft).
(4) That a "spotter" be designated when equipment is working near overhead lines. This person's responsibility is to observe safe working clearances around all overhead lines and to direct the operator accordingly.
(5) That warning cones be used as visible indicators of the 3 m (10 ft) safety zone when working near overhead power lines.

Informational Note: "Working near," for the purpose of this informative annex, is defined as working within a distance from any overhead power line that is less than the combined height or length of the lifting device plus the associated load length and the required minimum clearance distance [as stated in N.2(1)]. Required clearance is expressed as follows:
Required clearance = lift equipment height or length + load length + at least 3 m (10 ft)

(6) That the local responsible person be notified at least 24 hours before any work begins to allow time to identify voltages and clearances or to place the line in an electrically safe work condition.

N.3 Policy. All employees and contractors shall conform to the OPP. The first line of defense in preventing electrical contact accidents is to remain outside the limited approach boundary. Because most company and contractor employees are not qualified to determine the system voltage level, a qualified person shall be called to establish voltages and minimum clearances and take appropriate action to make the work zone safe.

N.4 Procedures.

N.4.1 General. Prior to the start of all operations where potential contact with overhead electrical systems is possible, the person in charge shall identify overhead lines or equipment, reference their location with respect to prominent physical features, or physically mark the area directly in front of the overhead lines with safety cones, survey tape, or other means. Electrical line location shall be discussed at a pre-work safety meeting of all employees on the job (through a job briefing). All company employees and contractors shall attend this meeting and require their employees to conform to electrical safety standards. New or transferred employees shall be informed of electrical hazards and proper procedures during orientations.

On construction projects, the contractor shall identify and reference all potential electrical hazards and document such actions with the on-site employers. The location of overhead electrical lines and equipment shall be conspicuously marked by the person in charge. New employees shall be informed of electrical hazards and of proper precautions and procedures.

Where there is potential for contact with overhead electrical systems, local area management shall be called to decide whether to place the line in an electrically safe work condition or to otherwise protect the line against accidental contact. Where there is a suspicion of lines with low clearance [height under 6 m (20 ft)], the local on-site electrical supervisor shall be notified to verify and take appropriate action.

All electrical contact incidents, including "near misses," shall be reported to the local area health and safety specialist.

N.4.2 Look Up and Live Flags. In order to prevent accidental contacts of overhead lines, all aerial lifts, cranes, boom trucks, service rigs, and similar equipment shall use look up and live flags. The flags are visual indicators that the equipment is currently being used or has been returned to its "stowed or cradled" position. The flags shall be yellow with black lettering and shall state in bold lettering "LOOK UP AND LIVE."

The procedure for the use of the flag follows.

(1) When the boom or lift is in its stowed or cradled position, the flag shall be located on the load hook or boom end.
(2) Prior to operation of the boom or lift, the operator of the equipment shall assess the work area to determine the location of all overhead lines and communicate this information to all crews on site. Once completed, the operator shall remove the flag from the load hook or boom and transfer the flag to the steering wheel of the vehicle. Once the flag is placed on the steering wheel, the operator can begin to operate the equipment.
(3) After successfully completing the work activity and returning the equipment to its stowed or cradled position, the operator shall return the flag to the load hook.
(4) The operator of the equipment is responsible for the placement of the look up and live flag.

N.4.3 High Risk Tasks.

N.4.3.1 Heavy Mobile Equipment. Prior to the start of each workday, a high-visibility marker (orange safety cones or other devices) shall be temporarily placed on the ground to mark the location of overhead wires. The supervisors shall discuss electrical safety with appropriate crew members at on-site tailgate safety talks. When working in the proximity of overhead lines, a spotter shall be positioned in a conspicuous location to direct movement and observe for contact with the overhead wires. The spotter, equipment operator, and all other employees working on the job location shall be alert for overhead wires and remain at least 3 m (10 ft) from the mobile equipment.

All mobile equipment shall display a warning decal regarding electrical contact. Independent truck drivers delivering materials to field locations shall be cautioned about overhead electrical lines before beginning work, and a properly trained on-site or contractor employee shall assist in the loading or off-loading operation. Trucks that have emptied their material shall not leave the work location until the boom, lift, or box is down and is safely secured.

N.4.3.2 Aerial Lifts, Cranes, and Boom Devices. Where there is potential for near operation or contact with overhead lines or equipment, work shall not begin until a safety meeting is conducted and appropriate steps are taken to identify, mark, and warn against accidental contact. The supervisor will review operations daily to ensure compliance.

Where the operator's visibility is impaired, a spotter shall guide the operator. Hand signals shall be used and clearly understood between the operator and spotter. When visual contact is impaired, the spotter and operator shall be in radio contact. Aerial lifts, cranes, and boom devices shall have appropriate warning decals and shall use warning cones or similar devices to indicate the location of overhead lines and identify the 3 m (10 ft) minimum safe working boundary.

N.4.3.3 Tree Work. Wires shall be treated as live and operating at high voltage until verified as otherwise by the local area on-site employer. The local maintenance organization or an approved electrical contractor shall remove branches touching wires before work begins. Limbs and branches shall not be dropped onto overhead wires. If limbs or branches fall across electrical wires, all work shall stop immediately and the local area maintenance organization is to be called. When climbing or working in trees, pruners shall try to position themselves so that the trunk or limbs are between their bodies and electrical wires. If possible, pruners shall not work with their backs toward electrical wires. An insulated bucket truck is the preferred method of pruning when climbing poses a greater threat of electrical contact. Personal protective equipment (PPE) shall be used while working on or near lines.

N.4.4 Underground Electrical Lines and Equipment. Before excavation starts and where there exists reasonable possibility of contacting electrical or utility lines or equipment, the local area supervision (or USA DIG organization, when appropriate) shall be called and a request is to be made for identifying/marking the line location(s).

When USA DIG is called, their representatives will need the following:

(1) Minimum of two working days' notice prior to start of work, name of county, name of city, name and number of street or highway marker, and nearest intersection
(2) Type of work
(3) Date and time work is to begin
(4) Caller's name, contractor/department name and address
(5) Telephone number for contact
(6) Special instructions

Utilities that do not belong to USA DIG must be contacted separately. USA DIG might not have a complete list of utility owners. Utilities that are discovered shall be marked before work begins. Supervisors shall periodically refer their location to all workers, including new employees, subject to exposure.

N.4.5 Vehicles with Loads in Excess of 4.25 m (14 ft) in Height. This policy requires that all vehicles with loads in excess of 4.25 m (14 ft) in height use specific procedures to maintain safe working clearances when in transit below overhead lines.

The specific procedures for moving loads in excess of 4.25 m (14 ft) in height or via routes with lower clearance heights are as follows:

(1) Prior to movement of any load in excess of 4.25 m (14 ft) in height, the local health and safety department, along with the local person in charge, shall be notified of the equipment move.
(2) An on-site electrician, electrical construction representative, or qualified electrical contractor should check the intended route to the next location before relocation.
(3) The new site is to be checked for overhead lines and clearances.
(4) Power lines and communication lines shall be noted, and extreme care used when traveling beneath the lines.
(5) The company moving the load or equipment will provide a driver responsible for measuring each load and ensuring each load is secured and transported in a safe manner.
(6) An on-site electrician, electrical construction representative, or qualified electrical contractor shall escort the first load to the new location, ensuring safe clearances, and a service company representative shall be responsible for subsequent loads to follow the same safe route.

If proper working clearances cannot be maintained, the job must be shut down until a safe route can be established or the necessary repairs or relocations have been completed to ensure that a safe working clearance has been achieved.

All work requiring movement of loads in excess of 4.25 m (14 ft) in height are required to begin only after a general work permit has been completed detailing all pertinent information about the move.

N.4.6 Emergency Response. If an overhead line falls or is contacted, the following precautions should be taken:

(1) Keep everyone at least 3 m (10 ft) away.
(2) Use flagging to protect motorists, spectators, and other individuals from fallen or low wires.
(3) Call the local area electrical department or electric utility immediately.
(4) Place barriers around the area.
(5) Do not attempt to move the wire(s).
(6) Do not touch anything that is touching the wire(s).
(7) Be alert to water or other conductors present.
(8) Crews shall have emergency numbers readily available. These numbers shall include local area electrical department, utility, police/fire, and medical assistance.
(9) If an individual becomes energized, DO NOT TOUCH the individual or anything in contact with the person. Call for emergency medical assistance and call the local utility immediately. If the individual is no longer in contact with the energized conductors, CPR, rescue breathing, or first aid should be administered immediately, but only by a trained person. It is safe to touch the victim once contact is broken or the source is known to be de-energized.
(10) Wires that contact vehicles or equipment will cause arcing, smoke, and possibly fire. Occupants should remain in the cab and wait for the local area electrical department or utility. If it becomes necessary to exit the vehicle, leap with both feet as far away from the vehicle as possible, without touching the equipment. Jumping free of the vehicle is the last resort.
(11) If operating the equipment and an overhead wire is contacted, stop the equipment immediately and, if safe to do so, jump free and clear of the equipment. Maintain your balance, keep your feet together and either shuffle or bunny hop away from the vehicle another 3 m (10 ft) or more. Do not return to the vehicle or allow anyone else for any reason to return to the vehicle until the local utility has removed the power line from the vehicle and has confirmed that the vehicle is no longer in contact with the overhead lines.

Informative Annex O Safety-Related Design Requirements

This informative annex is not a part of the requirements of this NFPA document but is included for informational purposes only.

O.1 Introduction. This informative annex addresses the responsibilities of the facility owner or manager or the employer having responsibility for facility ownership or operations management to perform a risk assessment during the design of electrical systems and installations.

O.1.1 This informative annex covers employee safety-related design concepts for electrical equipment and installations in workplaces covered by the scope of this standard. This informative annex discusses design considerations that have impact on the application of the safety-related work practices only.

O.1.2 This informative annex does not discuss specific design requirements. The facility owner or manager or the employer should choose design options that eliminate hazards or reduce risk and enhance the effectiveness of safety-related work practices.

O.2 General Design Considerations.

O.2.1 Employers, facility owners, and managers who have responsibility for facilities and installations having electrical energy as a potential hazard to employees and other personnel should ensure that electrical hazards risk assessments are performed during the design of electrical systems and installations.

O.2.2 Design option decisions should facilitate the ability to eliminate hazards or reduce risk by doing the following:

(1) Reducing the likelihood of exposure

(2) Reducing the magnitude or severity of exposure

(3) Enabling achievement of an electrically safe work condition

O.2.3 Incident Energy Reduction Methods. The following methods have proved to be effective in reducing incident energy:

(1) Zone-selective interlocking. A method that allows two or more circuit breakers to communicate with each other so that a short circuit or ground fault will be cleared by the breaker closest to the fault with no intentional delay. Clearing the fault in the shortest time aids in reducing the incident energy.

(2) Differential relaying. The concept of this protection method is that current flowing into protected equipment must equal the current out of the equipment. If these two currents are not equal, a fault must exist within the equipment, and the relaying can be set to operate for a fast interruption. Differential relaying uses current transformers located on the line and load sides of the protected equipment and fast acting relay.

(3) Energy-reducing maintenance switching with a local status indicator. An energy-reducing maintenance switch allows a worker to set a circuit breaker trip unit to operate faster while the worker is working within an arc flash boundary, as defined in NFPA 70E, and then to set the circuit breaker back to a normal setting after the work is complete.

O.2.4 Other Methods.

(1) Energy-reducing active arc flash mitigation system. This system can reduce the arcing duration by creating a low impedance current path, located within a controlled compartment, to cause the arcing fault to transfer to the new current path, while the upstream breaker clears the circuit. The system works without compromising existing selective coordination in the electrical distribution system.

(2) Arc flash relay. An arc flash relay typically uses light sensors to detect the light produced by an arc flash event. Once a certain level of light is detected the relay will issue a trip signal to an upstream overcurrent device.

(3) High-resistance grounding. A great majority of electrical faults are of the phase-to-ground type. High-resistance grounding will insert an impedance in the ground return path and will typically limit the fault current to 10 amperes and below (at 5 kV nominal or below), leaving insufficient fault energy and thereby helping reduce the arc flash hazard level. High-resistance grounding will not affect arc flash energy for line-to-line or line-to-line-to-line arcs.

(4) Current-limiting devices. Current-limiting protective devices reduce incident energy by clearing the fault faster and by reducing the current seen at the arc source. The energy reduction becomes effective for current above the current-limiting threshold of the current-limiting fuse or current limiting circuit breaker.

Informative Annex P Aligning Implementation of This Standard with Occupational Health and Safety Management Standards

This informative annex is not a part of the requirements of this NFPA document but is included for informational purposes only.

P.1 General. Injuries from electrical energy are a significant cause of occupational fatalities in the workplace in the United States. This standard specifies requirements unique to the hazards of electrical energy. By itself, however, this standard does not constitute a comprehensive and effective electrical safety program. The most effective application of the requirements of this standard can be achieved within the framework of a recognized health and safety management system standard. ANSI/AIHA Z10, *American National Standard for Occupational Health and Safety Management Systems*, provides comprehensive guidance on the elements of an effective health and safety management system and is one recognized standard. ANSI/AIHA Z10 is harmonized with other internationally recognized standards, including CAN/CSA Z1000, *Occupational Health and Safety Management*; ANSI/ISO 14001, *Environmental Management Systems - Requirements with Guidance for Use*; and BS OSHAS 18001, *Occupational Health and Safety Management Systems*. Some companies and other organizations have proprietary health and safety management systems that are aligned with the key elements of ANSI/AIHA Z10.

The most effective design and implementation of an electrical safety program can be achieved through a joint effort involving electrical subject matter experts and safety professionals knowledgeable about safety management systems.

Such collaboration can help ensure that proven safety management principles and practices applicable to any hazard in the workplace are appropriately incorporated into the electrical safety program.

This informative annex provides guidance on implementing this standard within the framework of ANSI/AIHA Z10 and other recognized or proprietary comprehensive occupational health and safety management system standards.

Index

15 16 17 6 5 4 3 Cou/W

Sequence of Events for the Standards Development Process

As soon as the current edition is published, a Standard is open for Public Input

Step 1: Input Stage

- Input accepted from the public or other committees for consideration to develop the First Draft
- Committee holds First Draft Meeting to revise Standard (23 weeks)
  Committee(s) with Correlating Committee (10 weeks)
- Committee ballots on First Draft (12 weeks)
  Committee(s) with Correlating Committee (11 weeks)
- Correlating Committee First Draft Meeting (9 weeks)
- Correlating Committee ballots on First Draft (5 weeks)
- First Draft Report posted

Step 2: Comment Stage

- Public Comments accepted on First Draft (10 weeks)
- If Standard does not receive Public Comments and the Committee does not wish to further revise the Standard, the Standard becomes a Consent Standard and is sent directly to the Standards Council for issuance
- Committee holds Second Draft Meeting (21 weeks)
  Committee(s) with Correlating Committee (7 weeks)
- Committee ballots on Second Draft (11 weeks)
  Committee(s) with Correlating Committee (10 weeks)
- Correlating Committee First Draft Meeting (9 weeks)
- Correlating Committee ballots on First Draft (8 weeks)
- Second Draft Report posted

Step 3: Association Technical Meeting

- Notice of Intent to Make a Motion (NITMAM) accepted (5 weeks)
- NITMAMs are reviewed and valid motions are certified for presentation at the Association Technical Meeting
- Consent Standard bypasses Association Technical Meeting and proceeds directly to the Standards Council for issuance
- NFPA membership meets each June at the Association Technical Meeting and acts on Standards with "Certified Amending Motions" (certified NITMAMs)
- Committee(s) and Panel(s) vote on any successful amendments to the Technical Committee Reports made by the NFPA membership at the Association Technical Meeting

Step 4: Council Appeals and Issuance of Standard

- Notification of intent to file an appeal to the Standards Council on Association action must be filed within 20 days of the Association Technical Meeting
- Standards Council decides, based on all evidence, whether or not to issue the Standards or to take other action

Committee Membership Classifications[1,2,3,4]

The following classifications apply to Committee members and represent their principal interest in the activity of the Committee.

1. M *Manufacturer:* A representative of a maker or marketer of a product, assembly, or system, or portion thereof, that is affected by the standard.
2. U *User:* A representative of an entity that is subject to the provisions of the standard or that voluntarily uses the standard.
3. IM *Installer/Maintainer:* A representative of an entity that is in the business of installing or maintaining a product, assembly, or system affected by the standard.
4. L *Labor:* A labor representative or employee concerned with safety in the workplace.
5. RT *Applied Research/Testing Laboratory:* A representative of an independent testing laboratory or independent applied research organization that promulgates and/or enforces standards.
6. E *Enforcing Authority:* A representative of an agency or an organization that promulgates and/or enforces standards.
7. I *Insurance:* A representative of an insurance company, broker, agent, bureau, or inspection agency.
8. C *Consumer:* A person who is or represents the ultimate purchaser of a product, system, or service affected by the standard, but who is not included in (2).
9. SE *Special Expert:* A person not representing (1) through (8) and who has special expertise in the scope of the standard or portion thereof.

NOTE 1: "Standard" connotes code, standard, recommended practice, or guide.

NOTE 2: A representative includes an employee.

NOTE 3: While these classifications will be used by the Standards Council to achieve a balance for Technical Committees, the Standards Council may determine that new classifications of member or unique interests need representation in order to foster the best possible Committee deliberations on any project. In this connection, the Standards Council may make such appointments as it deems appropriate in the public interest, such as the classification of "Utilities" in the National Electrical Code Committee.

NOTE 4: Representatives of subsidiaries of any group are generally considered to have the same classification as the parent organization.

Submitting Public Input / Public Comment through the Electronic Submission System (e-Submission):

As soon as the current edition is published, a Standard is open for Public Input.

Before accessing the e-Submission System, you must first sign-in at www.NFPA.org. *Note: You will be asked to sign-in or create a free online account with NFPA before using this system*:

a. Click in the gray Sign In box on the upper left side of the page. Once signed-in, you will see a red "Welcome" message in the top right corner.
b. Under the Codes and Standards heading, Click on the Document Information pages (List of Codes & Standards), and then select your document from the list or use one of the search features in the upper right gray box.

OR

a. Go directly to your specific document page by typing the convenient short link of www.nfpa.org/document#, (Example: NFPA 921 would be www.nfpa.org/921) Click in the gray Sign In box on the upper left side of the page. Once signed in, you will see a red "Welcome" message in the top right corner.

To begin your Public Input, select the link The next edition of this standard is now open for Public Input (formally "proposals") located on the Document Information tab, the Next Edition tab, or the right-hand Navigation bar. Alternatively, the Next Edition tab includes a link to Submit Public Input online

At this point, the NFPA Standards Development Site will open showing details for the document you have selected. This "Document Home" page site includes an explanatory introduction, information on the current document phase and closing date, a left-hand navigation panel that includes useful links, a document Table of Contents, and icons at the top you can click for Help when using the site. The Help icons and navigation panel will be visible except when you are actually in the process of creating a Public Input.

Once the First Draft Report becomes available there is a Public comment period during which anyone may submit a Public Comment on the First Draft. Any objections or further related changes to the content of the First Draft must be submitted at the Comment stage.

To submit a Public Comment you may access the e-Submission System utilizing the same steps as previous explained for the submission of Public Input.

For further information on submitting public input and public comments, go to: http://www.nfpa.org/publicinput

Other Resources available on the Doc Info Pages

Document information tab: Research current and previous edition information on a Standard

Next edition tab: Follow the committee's progress in the processing of a Standard in its next revision cycle.

Technical committee tab: View current committee member rosters or apply to a committee

Technical questions tab: For members and Public Sector Officials/AHJs to submit questions about codes and standards to NFPA staff. Our Technical Questions Service provides a convenient way to receive timely and consistent technical assistance when you need to know more about NFPA codes and standards relevant to your work. Responses are provided by NFPA staff on an informal basis.

Products/training tab: List of NFPA's publications and training available for purchase.

Community tab: Information and discussions about a Standard

Information on the NFPA Standards Development Process

I. Applicable Regulations. The primary rules governing the processing of NFPA standards (codes, standards, recommended practices, and guides) are the NFPA *Regulations Governing the Development of NFPA Standards (Regs)*. Other applicable rules include NFPA *Bylaws*, NFPA *Technical Meeting Convention Rules*, NFPA *Guide for the Conduct of Participants in the NFPA Standards Development Process*, and the NFPA *Regulations Governing Petitions to the Board of Directors from Decisions of the Standards Council*. Most of these rules and regulations are contained in the *NFPA Standards Directory*. For copies of the *Directory*, contact Codes and Standards Administration at NFPA Headquarters; all these documents are also available on the NFPA website at "www.nfpa.org."

The following is general information on the NFPA process. All participants, however, should refer to the actual rules and regulations for a full understanding of this process and for the criteria that govern participation.

II. Technical Committee Report. The Technical Committee Report is defined as "the Report of the responsible Committee(s), in accordance with the Regulations, in preparation of a new or revised NFPA Standard." The Technical Committee Report is in two parts and consists of the First Draft Report and the Second Draft Report. (See *Regs* at 1.4)

III. Step 1: First Draft Report. The First Draft Report is defined as "Part one of the Technical Committee Report, which documents the Input Stage." The First Draft Report consists of the First Draft, Public Input, Committee Input, Committee and Correlating Committee Statements, Correlating Input, Correlating Notes, and Ballot Statements. (See *Regs* at 4.2.5.2 and Section 4.3) Any objection to an action in the First Draft Report must be raised through the filing of an appropriate Comment for consideration in the Second Draft Report or the objection will be considered resolved. [See *Regs* at 4.3.1(b)]

IV. Step 2: Second Draft Report. The Second Draft Report is defined as "Part two of the Technical Committee Report, which documents the Comment Stage." The Second Draft Report consists of the Second Draft, Public Comments with corresponding Committee Actions and Committee Statements, Correlating Notes and their respective Committee Statements, Committee Comments, Correlating Revisions, and Ballot Statements. (See *Regs* at Section 4.2.5.2 and 4.4) The First Draft Report and the Second Draft Report together constitute the Technical Committee Report. Any outstanding objection following the Second Draft Report must be raised through an appropriate Amending Motion at the Association Technical Meeting or the objection will be considered resolved. [See *Regs* at 4.4.1(b)]

V. Step 3a: Action at Association Technical Meeting. Following the publication of the Second Draft Report, there is a period during which those wishing to make proper Amending Motions on the Technical Committee Reports must signal their intention by submitting a Notice of Intent to Make a Motion. (See *Regs* at 4.5.2) Standards that receive notice of proper Amending Motions (Certified Amending Motions) will be presented for action at the annual June Association Technical Meeting. At the meeting, the NFPA membership can consider and act on these Certified Amending Motions as well as Follow-up Amending Motions, that is, motions that become necessary as a result of a previous successful Amending Motion. (See 4.5.3.2 through 4.5.3.6 and Table1, Columns 1-3 of *Regs* for a summary of the available Amending Motions and who may make them.) Any outstanding objection following action at an Association Technical Meeting (and any further Technical Committee consideration following successful Amending Motions, see *Regs* at 4.5.3.7 through 4.6.5.3) must be raised through an appeal to the Standards Council or it will be considered to be resolved.

VI. Step 3b: Documents Forwarded Directly to the Council. Where no Notice of Intent to Make a Motion (NITMAM) is received and certified in accordance with the Technical Meeting Convention Rules, the standard is forwarded directly to the Standards Council for action on issuance. Objections are deemed to be resolved for these documents. (See *Regs* at 4.5.2.5)

VII. Step 4a: Council Appeals. Anyone can appeal to the Standards Council concerning procedural or substantive matters related to the development, content, or issuance of any document of the Association or on matters within the purview of the authority of the Council, as established by the *Bylaws* and as determined by the Board of Directors. Such appeals must be in written form and filed with the Secretary of the Standards Council (See *Regs* at 1.6). Time constraints for filing an appeal must be in accordance with 1.6.2 of the *Regs*. Objections are deemed to be resolved if not pursued at this level.

VIII. Step 4b: Document Issuance. The Standards Council is the issuer of all documents (see Article 8 of *Bylaws*). The Council acts on the issuance of a document presented for action at an Association Technical Meeting within 75 days from the date of the recommendation from the Association Technical Meeting, unless this period is extended by the Council (See *Regs at* 4.7.2). For documents forwarded directly to the Standards Council, the Council acts on the issuance of the document at its next scheduled meeting, or at such other meeting as the Council may determine (See *Regs* at 4.5.2.5 and 4.7.4).

IX. Petitions to the Board of Directors. The Standards Council has been delegated the responsibility for the administration of the codes and standards development process and the issuance of documents. However, where extraordinary circumstances requiring the intervention of the Board of Directors exist, the Board of Directors may take any action necessary to fulfill its obligations to preserve the integrity of the codes and standards development process and to protect the interests of the Association. The rules for petitioning the Board of Directors can be found in the *Regulations Governing Petitions to the Board of Directors from Decisions of the Standards Council* and in 1.7 of the *Regs*.

X. For More Information. The program for the Association Technical Meeting (as well as the NFPA website as information becomes available) should be consulted for the date on which each report scheduled for consideration at the meeting will be presented. For copies of the First Draft Report and Second Draft Report as well as more information on NFPA rules and for up-to-date information on schedules and deadlines for processing NFPA documents, check the NFPA website (www.nfpa.org/aboutthecodes) or contact NFPA Codes & Standards Administration at (617) 984-7246.